Zarina Zainuddin
Sahena Ferdosh
Nurul Hidayah Samsulrizal

Recent Advances in *Ficus* Research

Zarina Zainuddin
Sahena Ferdosh
Nurul Hidayah Samsulrizal

Recent Advances in *Ficus* Research

Noor Publishing

Imprint

Any brand names and product names mentioned in this book are subject to trademark, brand or patent protection and are trademarks or registered trademarks of their respective holders. The use of brand names, product names, common names, trade names, product descriptions etc. even without a particular marking in this work is in no way to be construed to mean that such names may be regarded as unrestricted in respect of trademark and brand protection legislation and could thus be used by anyone.

Cover image: www.ingimage.com

Publisher:
Noor Publishing
is a trademark of
Dodo Books Indian Ocean Ltd., member of the OmniScriptum S.R.L Publishing group
str. A.Russo 15, of. 61, Chisinau-2068, Republic of Moldova Europe
Printed at: see last page
ISBN: 978-620-3-86004-7

RECENT ADVANCES IN *Ficus*

RESEARCH

Zarina Zainuddin
Sahena Ferdosh
Nurul Hidayah Samsulrizal

TABLE OF CONTENT

PREFACE

The genus *Ficus* from the Moraceae family constitutes one of the largest family of Angiosperms with around 850 species distributed all over the world. *F. carica* (fig) is the most well-known *Ficus* species and other species include *F. deltoidea*, *F. racemosa*, *F. elastica* and *F. glomerata*. Ficus species has many benefits to human especially for fig where the benefits were mentioned in the Holy Quran, verse 1 to 4 of Surah At-Tin. Few hadiths also narrated the benefits of figs as said by Prophet Muhammad (PBUH). Abu ad-Darda' A.S narrated that the Prophet (PBUH) said: *"If I could say that a fruit was sent down from Heaven (to earth), I would say it is figs, because the Heaven's fruit has no stones. Eat it, as it cures hemorrhoids, and it is useful for treating gout" (Shahih al-Bukhari)*. The goodness of figs was also mentioned by Jewish Scriptures and the ancient Greeks. Hippocrates found that figs can cure inflammatory diseases such as inflammation in the respiratory tract and contain high laxative effect in increasing wisdom and mental health.

Realizing the importance of this genus, this book is being initiated by the Department of Plant Science, Kulliyyah of Science, International Islamic University Malaysia which compiles some of the research works conducted by the experts at the department and systematic reviews on *Ficus* sp. The chapters cover topics ranging from *Ficus* biology, diversity, and its origin (Chapter 1), effects of planting medium and type of cuttings towards root development and anatomy of *F. carica* (Chapter 2 & 3) and best agronomic practices focusing on nutrient requirement, fertilizer management and Bris soil improvement for fig's growth (Chapter 4 & 5). Chapter 6 and 7 provide comprehensive reviews on current trend in biotechnology and molecular biology of fig and various approaches for optimum *in vitro* propagation of *Ficus* sp. Research on genome size determination of *F. deltoidea* using flow cytometry method is being described in Chapter 8 while Chapter 9 highlights the potential of peat moss substitution with BRIS soil for the propagation of fig stem cutting. The last three chapters summarize the major pests and diseases in fig cultivation in Malaysia (Chapter 10), phytochemical screening methods, bioactive compounds, and extraction methods of common species of *Ficus* (Chapter 11) and pharmacological properties of the *Ficus* genus (Chapter 12). With the publication of this book, we hope that it will be invaluable to broad spectrum of readers including academicians, researchers, students, and those in commercial and pharmaceutical industries and will be source of reference for those who are interested on the current trend of *Ficus* research.

Editors

Zarina Zainuddin
Sahena Ferdosh
Nurul Hidayah Samsulrizal

An Overview of Fig (*Ficus carica* L.) Biology and Diversity

Halimaton Saaadiah, O. [1*]

[1] Department of Plant Science, Kulliyyah of Science, International Islamic University Malaysia, Jalan Sultan Ahmad Shah, Bandar Indera Mahkota, 25200,25200 Kuantan, Pahang, Malaysia.

[*]Corresponding Author: halimatonsaadiah@iium.edu.my

ABSTRACT

Fig (*Ficus carica* L.) is a member of Moraceae family, and it is the only species in the genus that is planted for fruit production. Fig is a hermaphrodite diploid that can be categorised with 4 types of fruit. Fig pollination is normally assisted through pollinator wasp known as *Blastophaga psenes* L. Major genetic erosion has posed a threat to fig germplasm over the last several decades. In order to overcome this, a few fig germplasm collections was established such as in North America, North Africa, Europe and Asia. Characterisation of fig cultivars are usually based on conventional methods which rely on phenotypic characterisation. This review describes the details involving fig biology as well as fig origin and diversity.

Keywords: Biology, Diversity, Fig, Pollination.

INTRODUCTION

Historically, fig is one among the most ancient, domesticated fruit. Fig gained its name from a Latin word derivation which is ficus. The name carica which derived from Caria, refers to an old area located in Asia Minor (Ferguson et al., 1990). Fig cultivation was expanded by the Carthaginians, Phoenicians and Greeks. It spread throughout the Mediterranean basin as a result of the Romans exploration (IBPGR, 1986; Roger, 2003). Later, fig cultivation was slowly introduced and expanded around other part of the remote area of the globe such as India, Australia, Arabia, China, the Americas and Ethiopia (Esclapon, 1976).

In various civilizations, fig is known by multiple names: caprificus (Romans), erineos (Greeks), dhokkar or ettine (Arabs), mgyz (Persians) and teb (Egyptians). The Arabs were responsible in promoting cultivation of fig which later become popular among the Mediterranean countries. They were known as figo de toco in Portuguese and caprificus or proficus in Italian (Lahbib, 1984). Although in general most of the member in *Ficus* genus thrive well around the tropical regions, fig could also thrive in subtropical climate (Ferguson et al., 1990). Agronomically, fig requires temperate and warm climates and has become one of the most common fruit tree planted around the Mediterranean region (Sahin, 1998). Fig cultivations are usually alongside with other trees such as almond, grapevines, pomegrates and olives (Aljane, 2011).

In Moraceae family, there are estimated around 60 genera which includes more than 1400 species. *Ficus* is a genus with over 700 species (Berg, 2003). Fig is one of the distinct *Ficus* species planted in a wide range of area from tropical and subtropical countries as the tree produces high commerial value of edible fruits (Irget et al., 2008). Fig can reach up to15-20 feet in height with multiple spreading branches and a single trunk less than 7 feet in diameter (Badgujar et al., 2014). The fruits of this species are drupelets that develop from the female ovaries contained in a closed syconium. Fig main pollinator is fig wasp known as *Blastophaga psenes* L. which made an opening (ostiole) in order to enter the fruit (Ferguson et al., 1990).

When the fruit is fresh, the skin is usually tender and thin and the colour of the fleshy wall varies either purple, pink, red, rose white or pale yellow. Fig becomes sweet and juicy upon maturity and may develop gum while immature. The size and number of seeds per fruit vary greatly with minimum of 30 and up to 1600 and the leaves resemble the shape of a hand (Poeaim et al., 2012). The fig trees wood can easily break as it is low in density. The interior of the branches is pithy. All broken plant structures produce gum, which can cause irritation when come in contact to human skin (Stover and Aradhya, 2008).

Fig is a hermaphrodite diploid species with total number of chromosome 26 which reproductively enable it to generate hermaphroditic or female flowers on different plants (Ferguson et al., 1990). This is one of the most prominent feature which could assist in classifying different types and cultivars of fig (García-Ruiz et al., 2013). Fig is made up of various cultivars with wide arrays of genetic variabilities that pose significant commercial values. The purpose of this paper is to provide an overview of fig biology and diversity covering its cultivation, germplasm diversity, utilisation and pollination process.

Production and Economic Importance

Total fig cultivation area surpasses 390,000 ha, with yearly fig production totaling 1,117,452 mt (FAO, 2015). The main fig producer with a total of 70% global production is Turkey followed closely by Tunisia, Iran, Egypt, Syria, Morocco and Algeria. Tunisia's annual fig production is 29,000 mt, accounting for 3% of global fig production (FAO, 2015). Spain is among the top fig producer in Europe with 23,285 mt which contributes 25% of European production and 3% of fig global fig production. Among the top fig producing European countries are Croatia, reece, Portugal, Italy, Albania, Germany and France (FAO, 2015).

The market demand for fresh and dried figs globally keep increasing, therefore it is important to continously maintain fig production worldwide. The dried fig is known to contain minerals, vitamins, sugars, carbohydrate, organic acids and phenolic compounds which is known to be higher than red wine and tea (Mawa et al. 2013). Both fresh and dried figs also contain high amounts of fiber and polyphenols. Parts of the fig's fruit, root and leaves are used in traditional medicine in treating gastrointestinal (colic, indigestion, loss of appetite and diarrhea), respiratory (sore throat, cough and bronchial issues) and cardiovascular disorder as well as as anti-inflammatory and antispasmodic remedy (Duke e tal., 2002). Fresh figs have a short shelf life, which presents a significant challenge in international marketing and commercial production. Both Mexico and Turkey have increased their total production for fig. Furthermore, with the improvement in cultural practises and the discovery of Calimyrna cv. in the Europe and United States as well as improved fresh-fruit postharvest, has widen up more opportunities for this crop to be further commercialised around the world (Flaishman et al., 2007).

Origin and Diversity Center

Smyrna figs are native to the Mediterranean and Middle East (Kjellberg and Valdeyron, 1984). Apart from its natural habitat, which is the Mediterranean regions, fig can rarely be found. (Nabli, 1989). Every year, fig trees may undergo spontaneous exchange of genetic material (Kjellberg et al., 1987). Alleles were not randomly distributed based on the enzymatic polymorphism, and their frequencies were dependent on the fluctuating environment, indicating the function of adaptation, according to a study using fig trees in southern France. (Kjellberg and Valdeyron, 1984).

The vast cultivation and spreading of fig tree throughout the world was as a result of human interest which represent fig tree affinity and adaptability to warm climates. Over the centuries, fig varieties were selected and multiplied through vegetative propagation (Storey, 1975; IBPGR, 1986). Fig diversity hotspots were regarded in Mediterranean countries such as Turkey, Spain, Algeria, Morocco and Tunusia. The primary gene pools of plant material that has traditionally been grown in different areas are known as microgenecenters (Mars, 2003).

Horticultural Classification and Reproductive Biology

The Eusyce subgenus includes figs, which are characterised by their unisexual inflorescence and gynodioecism. (Storey, 1975). The fig inflorescence differs from others in the genus because it has a syconium within that is implanted into the flower which bear the true fruits, tiny pedicellate druplets commonly referred to as the fig's "seed" (Storey, 1975; IBPGR, 1986). Morphologically and functionally, fig is gynodioecious. Trees may be divided into two categories based on the sex of the flowers. The first type is with syconia which has the short-styled pistillate flowers and staminate flowers surrounded around the inside of the ostiole cover, much of the inner wall of the caprifig. The domestic or female fig tree, on the other hand, bears solely long styled pistillate blooms on its syconia. Practically, only the female fig tree yields figs that are edible (Valdeyron, 1967; Storey, 1975).

According to Storey (1976), there are four catagories of figs:

(a) The most common, which does not require pollination thus producing infertile seeds. A good first crop (breba crop) will be produced followed by a successful second crop (as the main crop);

(b) The breba crop is produced by the San Pedro variety with parthenocarpy as the first crop. In order to reach maturity, pollination is required for the second crop (main crop). The production of the breba crop is the speciality of this type.

(c) The Smyrna variety yields ripe fruit with viable seeds and pollination requires caprifig pollen (for caprification). However, if pollination is not successful, the fruits would be aborted and dehisced off the trees.

(d) The Caprifig variety has pollen in its syconia flowers that can be used in pollinating Smyrna and San Pedro varieties.

POLLINATION

Pollination Mechanism

Fig plant and its pollinator wasp *Blastophaga psenes* L. have a symbiotic relationship, which results in an interesting pollination mechanism. The fig relies on the wasp to produce seeds, and the wasp spends its entire life cycle within the inflorescence or syconium. Male and female unisexual flowers are produced, but female flowers mature first. The female wasp lays eggs and pollinates female flowers within the syconium. Within the syconium, there are two types of female flowers: short styled and long styled. The wasp uses its ovipositor to enter the short-styled flowers and lay an egg in the ovary. As the developing wasp larvae feed on the ovary tissue, these short-styled flowers develop into galls. The female wasp pollinates these flowers with pollen collected from a male stage syconium because the style of the long-styled flower is longer than the ovipositor. As the wasp larvae grow, the syconia and its seed develop slowly. The male flowers of the syconium become matured when the wasps, both female and male, emerge from their galls within the syconium. The wasps mate within the syconium, and the males die after spending their entire life in this enclosure. Fertilized females collect pollen from male flowers, exit the male stage syconium, and transport the pollen to a female stage syconium via the ostiole. This entire process is known as caprification (Aljane and Ferchichi, 2007; Essid et al. 2015, 2017).

Sync003 development in some cultivars require pollination. Individual syconia's receptiveness can last up to two weeks and long-distance volatile substances are produced for several days in order to attract the fig wasps (Kjellberg et al., 1983; Kjellberg et al., 1988). Within this period, if pollination occurs fruit set and seed development would take place. Several wasps would visit most of the female syconium, which would result in pollination. After pollination, syconia would abruptly cease its receptiveness (Valizadeh et al., 1987; Khadari et al., 1995).

Substituting plant growth regulator sprays for caprification has yielded significant success. Among the many substances tested indole-3-butyric acid (IBA), 1-naphthaleneacetic acid (NAA), 2,4,5-Trichlorophenoxyacetic acid (2,4,5-T) and 4-chlorophenoxyacetic acid (4-CPA) were found to be most successful in inducing early maturing parthenocarpic figs (Crane and Blondcau, 1949; Crane and Brown, 1950). Fruit with parthenocarphy usually can reach the normal fruit size and have a desirable sugar content but being completely seedless thus lacking the crunch quality. These characteristics lead to the refusal on the use of Calimyrna figs in the baking industries in the United State. Crane (1952) later discovered that 2-3H-benzoxazolinone (BOA) induced parthenocarpy and the formation of drupelets with hollow, sclerified endocarps, but the mechanism is still unknown.

Pollination and Mutualism

Although dioecious, fig usually function as gynodioecious. Male and female flowers are produced by male trees (caprifigs). Female trees, on the other hand, only produce female flowers, upon pollination would produce fertile seeds. Strong mutualism is required between pollination and pollinating fig wasp (Herre et al., 2008). Caprifigs, also known as male figs, could generate three crop every year: profichi during summer, mammoni during autumn, and mamme during spring (Vidaud et al., 1997; Aljane and Ferchichi, 2007).

Farmers can induce the process of caprification, or it can happen naturally. Caprification could be triggered by the lack of fig wasps especially in cold regions, or lack of male flower during the time of female receptiveness (Aljane and Ferchichi, 2007; Esid et al., 2015, 2017). Pollen from the caprifig is used to pollinate Smyrna cultivars in early summer (Essid et al., 2015; 2017). About 15-20 female fig trees could be pollinated by a single male wasp. When the first wasps appear, picked profichi caprifigs are planted in the orchard. During the female fig's receptive period, caprifigs are replaced (Aljane and Nahdi, 2014; Essid et al., 2017). Fruit characteristics are highly dependant on the pollen's origin (Aksoy et al., 2003).

GENETIC RESOURCES

Germplasm Diversity and Conservation

The threat of genetic degradation has grown throughout the years and is one of the main issue in fig germplasm. Thus, in order to conserve and evaluate the fig germplasm resources, some countries have initiated several programmes. Lists of accessions of fig germplasm have been cited in North Africa, Europe and North America. Table 1, Table 2, and Table 3 list the reported fig germplasm collections. Turkey (Elbreyli Fig Research Institute), California (National Clonal Germplasm Repository (NCGR) Davis), Italy (Elbreyli Fig Research Institute), Turkmenistan (TES-PGR), and France (CBNM Porquerolles) have the most comprehensive germplasm collections. 154 different fig cultivars from France, Spain, Italy, Tunisia, North Africa, Greece, Turkey are included in the CBNM Porquerolles collection. Living plants produced from self-rooted cuttings are used to preserve germplasm. Only a few reports on in-situ or on-farm conservation and replanting were presented (Neffati and Mars, 1999). Despite the fact that the IBPGR (1986) indicated that seeds may and should be kept, as well as seed germination procedures were outlined (Storey, 1975; Jona and Gribaudo, 1991), there have been no reported outcomes on seed conservation. Certain geographical locations were chosen in effort to preserve and replant indigeneous fig cultivars.

The most ancient area for fig cultivation was identified in the southeastern Tunisian regions of Zammour and Douiret (Mars et al., 2008). As a result of the shifting environmental circumstances caused by climate change, there is a need to highlight on the importance of fig conservation through *in vitro*. Tissue culture is used to conserve genetic resource collections *in vitro* (IPGRI and CIHEAM, 2003). Cryopreservation and slow-growth approach are two *in vitro* conservation techniques being used (Engelmann, 2000).

For long-term storage, liquid nitrogen at 196°C is required which will fix both metabolic processes and cellular divisions. However, genetic instability poses a threat to stored plant materials (Panis et al., 2001). For cryopreservation of fig materials, two methods are used: the vitrification process as well as the traditional approach (Engelmann, 2000). Vitrification method is the most widely used method in plant cryopreservation method which employed tissue culture (Panis et al., 2001).The traditional protocol consists of two steps: plant materials are frozen using a gradual controlled cooling temperature followed by a quick immersion in liquid nitrogen. Cryoprotectant such as polyethylene glycol and dimethyl sulfoxide are also required, where water crystallisation within the cell and cell damage could be prevented (Panis et al., 2001).

In order to limit development and preserve the material, buds, meristems and shoot tips need to be cultured under cold conditions (Engelmann, 2000). The most effective method to preserve shoot tips through *in vitro* is the slow-growth technique which is the most commonly used method for somatic embryos and cell cultures. For root tips, it was found that no difference between slow-growth method and cryopreservation can be observed. Through adventitious bud development, plant regeneration from the root tip was achieved (Towill and Bajaj, 2002).

Table 1: List of fig accessions in European fig germplasm collection

Country	Locality and quntity of accessions	Accession	Origin	Citation
France	CBNM Porquerolles: 154	Abicou Boujassottenoire Col de dame noir Dauphine Longue d''aout Madeleine des deux sainsons Marseillaise Precoce ronde de Bordeux Sultana	Italy North Africa Spain Turkey France Greece France Tunisia	CBNMP (1979) Khadari et al. (1995) Vidaud et al. (1997)
Italy	Sezione Di Caserta : 442	Dottato Mattalona San Pietro	Italy	Grassi and Santonastaso (1998) Piga et al. (2008)
Spain	Badajoz : 85 Estremadure (SIFT) : 35	Alacantina Albatera Goen	Spain	Giraldo (2005) Giraldo et al. (2008)
Portugal	Alcobaca: 50 Faro (Algarve) : 60	Lampa Preta San Joao Branco	Portugal	Antunes et al. (2008)
Greece	Athens :16 Kalamata : 40 Houssa : 25	Tsapela Basanata Livano	Greece	Tsantili (1990) Lionakis (1996) Papadopulou et al. (2002)
Turkey	Adana : 35 Mersin : 40 Adana : 35 Albreyli (Fig Research Institute) : 314	Ipek Degirmen Istanbul Pamuk Patlican Kabak Sarilop (Calimyrna)	Turkey	Aksoy (1995) Bostan et al. (1998) Karadeniz (2008)

Table 2: List of fig accessions in Asian fig germplasm collection

Country	Locality and quntity of accessions	Accession	Origin	Citation
Albania	Tirana : 21	Cingell Tivaras Stambolli	Albania	Koka (2008)
Georgia	Abkhazia : 120			Petrova and Voronova (1984)
Serbia	Trebinje : 40	Sargulja Tenica Zimmica	Serbia	Kulina et al. (2002)
Slovenia	Ljubljana : 4	Bilicia Dobrica Zimica	Slovenia	Zigo and Stampar (2002)
China	Xuchang (Institute Forestry Henan) : 21	Branswisck Dauphine He-18 Masui	China	Jianye et al. (1997)
Syria	Idleb : 74	Abidi Gahzleni Sahli	Syria	Al Ibrahim (2000) Al Ibrahim and Bari (2006)
Turkmenistan	TES-PGR : 180			Levina (1984)

Table 3: List of fig accessions in North American and North African fig germplasm collection

Country	Locality and quntity of accessions	Accession	Origin	Citation
United States of America	NCGR Davis California : 190	Panachee jaune verte Brunswick Violette de Bordeaux	California France	GRIN (1990) Vidaud et al. (1997) Simon (2002) Stover and Aradhya (2008)
Algeria	Sazione di Caserta : 58	Alekake Tabouharchaout Azendjar Bakkar Taanimt Tabarant Tamaghroust (parthenocarpic)	Algeria	IBPGR (1986) Mazri-Kartout and Aid-Houchi (2001)
Morocco	Ain Taoujdate (UNRA Meknes) : 72	Nabout Naoukali Ghoddane Hafer El-Bghal Bousbati Fassi	Morocco	Messaoudi (2003) Oukabli et al. (2003)

		Filalia		Khadari et al.
		Embar El-Khel		(2004)
		Abacor Blanca		
		Blanca	Exotic	
		Cuello Damo		
		El Quolti Trojano		
		Lebied		
		Kadota		
Tunisia	Tataouine (Institute of	Zidi	Tunisia	Aljane and
	Arid Regions) : 139	Soltani		Ferchichi (2009)
	(119 female figs and 20	Temri		Aljane and
	male figs)	Wedlani		Ferchichi (2010)
		Magouli		Aljane (2011)
		Sawoudi		Aljane et al.
		Makhbech		(2012)
		Bouholi		Aljane (2016)
		Wahchi		Essid et al.
		Bouharrag		(2017)
		Magouli		
		Jrani		
		Asafri	Egypt	
		Limi	Algeria	
		Tebassi		
		Masria		
		Zazairia		

CONCLUSION AND FUTURE PROSPECTS

Towards improvement of fig *ex situ* conservation, information based on morphological as well as molecular data can be used. Presently, many countries experienced fig genetic erosion threat particularly involving Smyrna cultivar, thus improvement approaches taken in managing fig genetic resources should be emphasised. In many parts of the world, significant fig genetic diversity are available and can be used to select new fig cultivar that can adapt in arid and semi-arid environments. Data from quantitative variation analysis, particularly from elucidating its probable genetic basis, are so important for breeding programmes.

REFERENCES

Aksoy, U. (1995). Present status and future prospects of underutilized fruit production in Turkey. *Cah. Opt. Mediterr.* 13:97–107.

Aksoy, U., Balci, B. and Can, H. Z. (2003). Some significant results of the research-work in Turkey on fig. *Acta Hort.* 606:173–181.

Aljane, F. (2011). *Characterization, evaluation and selection of local fig tree (Ficus carica L,) accessions in Tunisia.* Doctoral thesis, Faculty of Sciences of Tunis, Tunisia.

Aljane, F. (2016). Analysis of genetic diversity in Tunisian fig (*Ficus carica* L.) germplasm bank revealed by RAPD markers and morphological characters. *Eur. J. Sci. Res.* 142(2):172–192.

Aljane, F. and Ferchichi, A. (2007). Characterisation and evaluation of six caprifig (*Ficus carica* L.) cultivars in Tunisia. *Plant Genet. Res. Newsletter* 151:22–26.

Aljane, F. and Ferchichi, A. (2009) Assessment of genetic diversity among some southern Tunisian fig (*Ficus carica* L.) cultivars based on morphological descriptors. *Jordan J. Agric. Sci.* 5(1):1–16 .

Aljane, F. and Ferchichi, A. (2010). Assessment of genetic diversity of Tunisian fig (*Ficus carica* L.) cultivars using morphological and chemical characters. *Acta Bot. Gallalica* 157(1):171–182.

Aljane, F. and Nahdi, S. (2014). Propagation of some local fig (*Ficus carica* L.) cultivars by hardwood cuttings under the field conditions in Tunisia. *Int. Sch. Res. Not.* 1–5.

Aljane, F., Nahdi, S. and Essid, A. (2012). Genetic diversity of some accessions of Tunisian fig tree (*Ficus carica* L.) based in morphological and chemical traits. *J. Natl. Prod. Plant Res.* 2(3):350–359.

Antunes, M. D. C., Costa, P. and Migel, M. G. (2008). The effect of postharvest treatments with sodium bicarbonate or acetic acid on storage ability and quality of fig fruit. *Acta Hortic.* 798:279–284.

Berg, C. C. (2003). Flora malesiana precursor for the treatment of Moraceae 1: the main subdivision of *Ficus*: the subgenera. *Blumea* 48:167–178.

Bostan, S. Z., Islam, A. and Aygün, A. (1998). A study pomological characteristics of local fig cultivars in Northern Turkey. *Acta Hort.* 480:71–73.

CBNMP. (1979). National Mediterranean Botanical Conservatory of Porquerolles, France, Island of Porquerolles. Retrieved from http://www.cbnmed.fr/pres/index.php

Crane, J. C. (1952). Ovary-wall development as influenced by growth-regulators. *Bot. Gaz.* 114102-107.

Crane, J. C. and Blondeau, R. (1949). The use of growth-regulating chemicals to induce parthenocarpic fruit in the Calimyrna fig. *Plant Physiol.* 24:44-54.

Crane, J. C. and Brown, J. B. (1950). Growth of the fig fruit, *Ficus carica* var. Mission. *Proc. Am. Soc. Hort. Sci.* 56:93-97.

De Masi, L., Cipollaro, M. and Di Bernardo, G. (2003). Clonal selection and molecular characterization by RAPD analysis of the Fig (*Ficus carica* L) ''Dottato'' and ''Bianco del Cilento'' cultivars in Italy. *Acta Hortic.* 605:65–68.

Demiralay, A., Yalçin-Mendi, Y., Aka-Kaçar, Y. and Çetiner, S. (1998). In vitro propagation of *Ficus carica* L. var. Bursa Siyahi through meristem culture. *Acta Hortic* 480:165–167.

Duke, J. A., Bugenschutz-godwin, M. J., Du collier, J. and Duke, P. K. (2002). *Hand Book of Medicinal Herbs*, CRC Press, 2nd edition. Boca Raton, Florida, USA.

El Rayes, R. (1996). *Country report of Syria*. In : Galan Sauco V. (ed.), Proc. First MESFIN Plant Genetic Resources Meeting, Tenerife, Spain: 2 - 4 October 1995: 105- 114.

Engelmann, F. (2000). *Importance of cryopreservation for the conservation of plant genetic resources*. In: Engelmann F, Takagi H (eds) Cryopreservation of tropical plant germplasm. Current research progress and application. IPGRI, Rome, pp 8–20.

Esclapon, R. G. (1976). *The fig tree and its future in South France*. Arboriculture Fruitière, France, pp. 266.

Essid, A., Aljane, F. and Ferchichi, A. (2015). Analysis of genetic diversity of Tunisian caprifig (*Ficus carica* L.) accessions using simple sequence repeat (SSR) markers. *Hered.* 152(1):1–7.

Essid, A., Aljane, F. and Ferchichi, A. (2017). Morphological characterization and pollen evaluation of some Tunisian *ex situ* planted caprifig (*Ficus carica* L.) ecotypes. *S Afr. J. Bot.* 111:134–143.

FAO (2015). *The FAO statistical database-agriculture*. Food Agriculture Organization. Retrieved from http://faostat.fao.org. Accessed on Aug 2017

Ferguson, L., Michailides, J. T. and Shorey, H. H. (1990). The California fig industry. *Hort. Rev.* 12:409–490.

Flaishman, M., Rodov, V. and Stover, E. (2007). *The fig: botany, horticulture and breeding*. In: Janick J (ed) Horticultural reviews, vol 34. Wiley, Hoboken, NJ, USA, pp 113–197.

Galderisi, U., Cipollaro, M. and Di Bernardo, G. (1999). Identification of the edible fig "Bianco Del Cilento" by random amplified polymorphic DNA analysis. *Hortic. Sci.* 34(7):1263–1265.

García-Ruiz, M. T., Mendoza-Castillo, V. M., Valadez-Moctezuma, E. and Muratalla-Lúa, A. (2013). Initial assessment of natural diversity in Mexican fig landraces. *Genet. Mol. Res.* Sep 23;12(3):3931-3943.

Ghorbel, A., Kchouk, M. L., Fnayou-Ben Salem, A., Mais, M. and A. Rhouma, (1996). *Morpho-qualitative descriptions in ex situ collections of date palm, fig and pomegranate*. In: Galan Sauco V. (ed.), Proc.

First MESFIN Plant Genetic Resources Meeting, Tenerife, Spain: 2 - 4 October pp: 125 - 140.

Giraldo, E. (2005). *Morphological and molecular characterisation of fig (Ficus carica L.) varieties*. Ph.D. thesis, University of Extremadura, Spain.

Giraldo, E., Lopez Corrales, M. and Hormaza, J. I. (2008). Selection of morphological quantitative variables in fig characterization. *Acta Hortic*. 798:103–108.

Grassi, G. And Santonastaso, M. (1998). The fig growing in Italy: the present state and problems. *Acta Hortic*. 480:31–35.

Gregoriou, C. (1995). Cultivation of fig (*Ficus carica*), loquat (*Eriobotrya japonica*), Japanese persimmon (*Diospyros kaki*), pomegranate (*Punica granatum*) and barbary fig (*Opuntia ficus indica*) in Cyprus. *Cah. Options Méditer*. Vol. 13, 1995 : 9 - 12.

GRIN (1990). Germplasm resources information network. United States Department of Agriculture, USA, Washington.

Haelterman, R. M. and Docampo, D. M. (1994). *In vitro* propagation of mosaic-free fig (*Ficus carica* L.) cultivars using thermotherapy and shoot-tip cultures. *Revista de Investigaciones Agrarias,* 25(3): 15 - 22.

Herre, E. A., Jander, K. C. and Machado, C. A. (2008). Evolutionary ecology of figs and their associates: recent progress and outstanding puzzles. *Ann. Rev. Ecol. Evol. Syst*. 39:439–458.

IBPGR (1986). *Ficus carica* L. In: Genetic resources of tropical and sub-tropical fruits and nuts. Rome.

Ikegami, H., Nogata, H., Inoue, Y., Himeno, S., Yakushiji, H., Hirata, C., Hirashima, K., Mori, M., Awamura, M. and Nakahara, T. (2013). Expression of FcFT1, a FLOWERING LOCUS T-like gene, is regulated by light and associated with inflorescence differentiation in fig (*Ficus carica* L.). *BMC Plant Biology* 13: 216–227.

IPGRI and CIHEAM (2003). *Descriptors for fig (Ficus carica L.)* International Plant Genetic Resources Institute (IPGRI). Rome Italy: International Center for Advanced Mediterranean Agronomic Studies (CIHEAM), Paris, France.

Irget, M. E., Aksoy,U., Okur, B., Ongun, A. R. and Tepecik, M. (2008). Effect of calcium based fertilization on dried fig (*Ficus carica* L. cv. Sarılop) yield and quality. *Scientia Horticulturae* 118:308.

Jianye, C., Yuxia, N. and Zilan, Z. (1997). Observation on biological characteristics of fig (*Ficus carica* L.). *J. Fruit Sci*. 14(1):16–20.

Jona, B. and Grihaudo, I. (1991). *Ficus* spp. In: Baja YPS (ed) Biotechnology in agriculture and forestry, vol 16. Springer, Berlin, pp. 76–93.

Karadeniz, T. (2008). Clonal selection in "Patlican" at Black Sea region of Turkey. *Acta Hortic*. 798:135–138.

Khadari, B., Lashermes, P. and Kjellberg, F. (1995). RAPD fingerprints for identification and genetic characterization of fig (*Ficus carica* L.) genotypes. *J. Genet. Breed*. 49:77–86.

Khadari, B., Oukabli, A. and Ater, M. (2004). Molecular characterization of Moroccan fig germplasm using inter simple sequence repeat and simple sequence repeat markers to establish a reference collection. *Hortic Sci*. 40:29–32.

Kjellberg, F. A., Aljibouri, G. and Valdeyron, G. (1983). Recent observation on fig tree pollination. *Fruits* 38(7-8):567-569.

Kjellberg, F., Doumesche, B. and Bronstein, J. L. (1988).: Longevity of a fig wasp (*Blastophaga psenes* L.). *Ecology* 91 (2): 117-122.

Kjellberg F, Gouyon P. H. and Ibrahim, M. (1987). The stability of the symbiosis between dioecious figs and their pollinators: a study of *Ficus carica* L. and *Blastophaga psenes* L. *Evolution*. 41:653–660.

Kjellberg, F. and Valdeyron, G. (1984). The pollination of fig tree (*Ficus carica* L.) and its control in horticulture. *Acta Oecologica* 5(4):407–412.

Koka, T. (2008). Fig germplasm conservation in Albania. *Acta Horticultural* 798:77–80.

Kuden, A. B. (1996). *Plant genetic resources of fig*. In: Galan Sauco V. (ed.), Proc. First MESFIN Plant Genetic Resources Meeting, Tenerife, Spain: 2 - 4 October 1995 : 323- 330.

Kulina, M., Djurdjic, Z. and Vico,. G (2002). Pomological traits of some once-bearing figs in the area of Trebinjie. *Acta Agric. Serbica* 7(13):9–15.

Lahbib, T. (1984). *Pomological study of fig varieties (Ficus carica L.) listed in the Tunisian Sahel.* Final Year Thesis, National Agronomic Institute of Tunis, Tunis, Tunisia.

Levina, E. K. (1984). A study of olive, fig and persimmon at the Turkmen experimental station of the VIR Sbornik Nauchnykh Trudov. *Prikladniui Botanike Genetike I Selektsii* 83:41–45.

Lionakis, S. M. (1996). *Genetic resources of plant grown in Greece and included in the MESFIN network.* In: Galan SV (ed) In: Proceedings of International MESFIN plant genetic resources meeting, Tenerife, Spain, p 25.

Mars, M. (2003). Fig (*Ficus carica.*L.) genetic resources and breeding. *Acta Hortic.* 605:19–26.

Mars, M., Chatti, K. and Saddoud, O. (2008) Fig cultivation and genetic resources in Tunisia. An overview. *Acta Hortic.* 798:27–32.

Mawa, S., Husain, K. and Jantan, I. (2013). *Ficus carica* L. (Moraceae): Phytochemistry, traditional uses and biological activities. *Evidence-Based Complementary and Alternative Medicine* 2013:1-8.

Mazri-Kartout, C. and Aid-Houchi, A. (2001). Contribution to the characterisation of three varietes of fig in Fréha Wilaya de Tizi-Ouzou. Agronomic Research. *INRA d'Algérie. Rev Semes* 14:53–63.

Messaoudi, Z. (2003). Propagation of five fig (*Ficus carica.*L.) varieties under field conditions *Acta Hortic.* 605:103–106.

Mori, K., Shirasawa, K., Nogata, H., Hirata, C., Tashiro, K., Habu, T., Kim, S., Himeno, S., Kuhara, S. and Ikegami, H. (2017). Identification of RAN1 orthologue associated with sex determination through whole genome sequencing analysis in fig (*Ficus carica* L.). *Sci. Rep.* 7:41124.

Neffati, M. and Mars, M. (1999*). Towards a participatory in situ conservation of plant genetic resources in the arid regions of Tunisia.* International Workshop on Strategies and technologies for Conservation and Sustainable Use of Biodiversity in Large Asian and North African landscapes, Marsa Matrouh, Egypt, March 13 - 18, 1999.

Oukabli, A., Khadari, B. and Roger, J. P. (2003). Genetic variability in Moroccan fig cultivars (*Ficus carica* L.) based on morphological and pomological data. *Acta Hortic.* 605:311–318.

Özen, M., Kocataş, H., Çobanoğlu, F., Ertan, B., Kösoğlu, İ., Tan, N., Şahin, B., Belge, A., Konak, R., Aksoy, U. and Gülşen, O. (2017). Mutation breeding studies of fig. *Acta Horticulturae* 1173: 93-98.

Panis, B., Swennen, R. and Engelmann, F. (2001.) Cryopreservation of plant germplasm. *Acta Hortic.* 560:79–86.

Papadopoulou, K., Ehaliotis, C. and Tourna, M. (2002). Genetic relatedness among dioecious *Ficus carica* L. Cultivars by random amplified polymorphic DNA analysis, and evaluation of agronomic and morphological characters. *Genetica* 114:183–194.

Petrova, E. F. and Voronova, O. G. (1984). Biological characteristics of figs in Abkhazia. *Naucho Tekhnich eskii Byulleon* 141:163–166.

Piga, A., Del Caro, A. and Milella, G. (2008). HPLC analysis of polyphenols in peel and pulp of fresh figs. *Acta Hortic.* 798:301–306.

Poeaim, A., Poraha, R., Chontong, R., Jantima, O. and Pongjaroenkit, S. (2012). *Callus Induction and Cell Suspension Cultures of Khao Pong Krai Rice (Oryza sativa L.)* In The 9th National Kasetsart University Kamphaeng Saen Conference. Nakhon Pathom, Thailand: Kasetsart University Kamphaeng Saen Campus. pp. 2182-2188.

Roger, J. P. (2003). The origin of fruit trees. National Mediterranean Botanical Conservatory of Porquerolle, Antenne Provence-Alpes, Côte d'Azur, France.

Sahin, N. (1998). Fig adaptation studies in Western Turkey. *Acta Hortic.* 480:61–70.

Shamkant, B. Badgujar, S. B., Patel, V. V., Bandivdekar, A. H. and Mahajan, R. T. (2014). Traditional uses, phytochemistry and pharmacology of *Ficus carica*: A review. *Pharmaceutical Biology* 52:11, 1487–1503.

Simon, C. J. (2002). Fabulous figs featured in California collection. *Agric. Res.*3:22-35.

Soliman, H. I. A. and Abd Alhady, M. R. A. (2017). Evaluation of salt tolerance ability in some fig (*Ficus carica* L.) cultivars using tissue culture technique. *J. App. Biol. Biotechnol.* 5(6):29–39.

Storey, W. B. (1975). *Figs*. In: Janick J, Moore JN (eds) Advances and fruit breeding Indiana. Purdue University Press, West Lafayette, Indiana, pp 568–589.

Storey, W. B. (1976). *Fig Ficus carica* (Moraceae). In: Simmonds NW (ed) Evolution of crop plants. Longman, London, pp 205–208.

Stover, E. and Aradhya, M. (2008). Fig genetic resources and research at the US national Clonal Germplasm repository in Davis, California. *Acta Hortic.* 798:57–68.

Shukranul, M., Khairana H. and Ibrahim J. (2013). Ficus carica L. (Moraceae): Phytovhemistry, traditional uses and biological activities. *Evidence-based Complementary and Alternative Medicine.* 2013:974-988.

Taha, R. A., Mustafa, N. S. and Hassan, S. A. (2013). Protocol for micropropagation of two *Ficus carica* cultivars. *World J. Agric. Sci.* 9(5):383–388.

Towill, L. E. and Bajaj, Y. P. S. (2002). *Biotechnology in agriculture and forestry*. Cryopreservation of plant germplasm II. Springer, USA

Trabut. L. (1922). The origins of fig trees. *Rev. Bot. Agr. Colon.*, 2: 396–398.

Tsantili, E. (1990). Changes during development of "Tsapela" fig fruits. *Sci Hortic.* 44:227–234.

Valdeyron, G. (1967). genetic system of the fig tree, Ficus carica L. Trial of evolutionary interpretation. Annales de l'INA, Tome 5, Paris, pp. 167.

Valizadeh, M., Valdeyron, G., Kjellberg, F. and Ibrahim, M. (1987). Gene flow in the common fig, Ficus carica: pollen dispersal in a dense stand. *Acta Ecologica Plantarum*, 8: 143–154.

Vidaud, J., Baccaunaud, M., Caraglio, Y., Hutin, C. and Roger, J. P. (1997). *Fig tree*. Technical and Interprofessional Centre for Fruits and Vegetables Paris, France.

Zigo, A. and Stampar, F. (2002). *Characterization of isozymes variation in common fig (Ficus carica L.).* Research Report. University of Ljubljana, Biotechnical Faculty, Institute of Fruit Growing, Viticulture and Vegetable Growing, Slovenia.

Root Development of *Ficus carica* cv. BTM6 as Affected by Planting Medium and Type of Cutting

Rozilawati Shahari, Che Nurul Aini Che Amri*, Mohd Syufiudin Shamsuddin and Nur Shuhada Tajuddin

¹Department of Plant Science, Kulliyyah of Science, International Islamic University Malaysia, Jalan Sultan Ahmad Shah, Bandar Indera Mahkota, 25200, Kuantan, Pahang, Malaysia.
**Corresponding author: <u>chenurulainicheamri@iium.edu.my</u>*

ABSTRACT

Fig (*Ficus carica* L.) is a member of family Moraceae that originated from the Mediterranean region. *Ficus carica* can easily be propagated by stem cutting. However, there is a lack of information on their root development. Thus, this experiment aims to study the root development of *Ficus carica* cv. BTM6 as affected by different planting medium and type of cutting. The research was conducted at the Glasshouse and Nursery Complex (GNC), International Islamic University Malaysia (IIUM) Kuantan Campus. Three mixtures of planting medium and two types of stem cutting were tested. The experiment was conducted with five replications in a split-plot design. The type of media was used as the main plot and cutting as a sub-plot. Plants were harvested 45 days after planting. Number of newly emerged root, primary root, secondary root and tertiary root were recorded. The root anatomical structure was prepared through a series of histological work. Samples were observed and analyzed using Leica LAS EZ software. Data such as root length, root diameter, cortex length, vascular bundle were observed and annotated. Based on this study, there were significant effects of planting medium type on the number of newly emerged roots. However, there was no significant effect of planting medium on primary, secondary, and tertiary roots. In terms of cutting type, there were no significant differences among all parameters measured. For anatomical structure, planting medium and type of cutting influenced the growth of *Ficus carica* cv. BTM6 root. In conclusion, this study showed that semi-hardwood cutting gave higher performance when grown in a combination of 50% peat and 50% perlite. Thus, the mixture of 50% peat: 50% perlite and semi-hardwood cutting was recommended to be used in the propagation of *Ficus carica* cv. BTM6.

Keywords: Cutting, *Ficus carica*, medium, root.

INTRODUCTION

Ficus carica L. is an important member of the genus *Ficus* from family Moraceae with more than 1400 species and 40 genera. It is ordinarily deciduous and commonly referred to as "fig". The common fig is originally from southwest Asia and the eastern Mediterranean region and is one of the first plants grown by people (Mawa et al., 2013). Fig can grow from 10 to 30 feet (California Rare Fruit Growers, 2016) and have different common names such as Anjir (India), Brevo (Spanish), and Figo (Italian) (Flaishman et al., 2008). Generally, there are four types of fig trees, including Caprifigs, Smyrna, San Pedro and common fig. Caprifig is an inedible type consisting of both male and female parts, Smyrna consists of only female flowers and must be pollinated to produce fruit, San Pedro is a breba crop which does not require pollination except for the main crop, while the common fig is completely parthenocarpic and undergoes self-pollination (Lawrence, 2009).

Ali et al. (2013) stated that the edible fig cultivars are generally propagated by stem cuttings. According to Acosta et al. (2009), propagation via stem cutting is a common method for the cultivation of various kinds of ornamental plants. The type of cutting used is one of the factor that could lead to higher survival and rooting percentage (Hassanein, 2013). In addition, at least one node (point where the buds form) must be placed below the surface of the propagation media (Relf & Ball, 2009).

According to Seago Jr & Fernando (2013), root is the vital part of vascular plants that provides anchorage to the upper part structure which also help absorbs dissolved minerals and water in order to facilitate growth and development of plant. The formation of root system is crucial, especially during the cutting propagation. The plant survival depends on carbohydrates content in the stem cutting. Larsen (1997) that there is no root system in cutting that could help water absorption. However, a small portion of water can be absorbed at the base of unrooted cutting which can only keep the cutting alive for a short period of time. Preti & Giadrossich (2009) stated that the distribution of root of the transplanted plants is due to soil condition of the plant itself. The planting medium which surround and support rooting area will determine the formation of root either it will emerge or not (Okanlawon et al., 2016). The root apical meristem (RAM) plays an important role in the growth and development of root formation during the early growth stage (Euzébio de Souza et al., 2015). Inadequate water and nutrient can also result in different growth ratio between root and the upper vegetative parts (Paez-Garcia et al., 2015). Dubrovsky and North (2002) reported that the elongation and meristem zone will decreases if the rooting medium is gradually dry. Thus, this study was conducted to study root development of Ficus *carica* cv. BTM6 as affected by different types of planting medium and stem cutting.

MATERIALS AND METHODS

Field experiment

The research was conducted at the Glasshouse and Nursery Complex (GNC), International Islamic University Malaysia (IIUM) Kuantan Campus. One year old *F. carica* cv. Brown Turkey Modified-6 (BTM6) was used as the mother plant. A total of 540 stem cuttings was used and each cutting was transplanted into a 6 cm x 9 cm polyethylene bag and placed under glasshouse. The experiment was conducted with five replications in a split plot design. There were 18 experimental units per treatment per block in this experiment where two factors were used, namely media type and stem cutting. Three mixtures of medium planting and two types of stem cutting were tested (Table 1). The type of media was used as the main plot and type of cutting as a sub-plot. The leafless cuttings were prepared with a similar internode length between 2.5 cm and 3.5 cm. Plants were harvested 45 days after planting. Data such as number of newly emerge root, number of primary root, number of tertiary root and number of tertiary root were recorded. Parts of the roots samples were collected and prepared for histological work. Analysis of variance (ANOVA) was used to test the statistical significance between experimental treatments. Multiple mean comparisons were performed using the Duncan New Multiple Range Test (DNMRT) for mean comparison. These statistical tests were performed by using open-source Statistical Analysis Software (SAS) version 9.4.

Table 1: Type of planting medium and stem cutting used in the experiment

Treatment	Planting Medium	Type of cutting	Treatment combination
T1	M1 (75% topsoil: 25% sand)	C1 (Semi-hardwood)	M1C1
T2		C2 (Hardwood)	M1C2
T3	M2 (33.33% topsoil: 33.33% sawdust : 33.33% sand)	C1 (Semi-hardwood)	M2C1
T4		C2 (Hardwood)	M2C2
T5	M3 (50% peat: 50% perlite)	C1 (Semi-hardwood)	M3C1
T6		C2 (Hardwood)	M3C2

Histological process

Root samples were fixed in the formalin-aceto-alcohol (FAA) solution right after harvesting for at least a week. The roots samples were washed by fixing the samples in 70% alcohol solution for at least 2 hours. The samples were then cut and pressed between two corks. Next, the samples were then carefully placed and adjusted to the sliding microtome to avoid tilting samples. The sectioning process of the samples was carried out by meticulously moving the sliding microtome. Once the tissue begins to appear, the samples were gently collected using the end of the hair paintbrush and placed in distilled water. The process of sectioning was repeated several times to obtain the finest tissue samples. The samples were then taken to the fume hood. 2 to 3 drops of chlorox were added to the tissue sample solution for 10 minutes. Next, the samples were washed 3 times with distilled water before the samples were soaked with 3 to 5 drops of methylene blue for 5 minutes. The solutions were again washed with distilled water 2 times to remove excessive colour. After that, 2 to 3 drops of safranin were added to the tissue sample solution for 5 minutes. The samples were washed again with distilled water for 2 times to remove any excessive colour. The following step was a process of dehydration using a series of alcohol concentrations. The tissue samples were soaked in 50%, 70%, 90% and 100% alcohol for 5 minutes at each alcohol concentration. The tissue samples were then moved into the slide using the end of the hair paintbrush. At the end, the samples were gently mounted using the euparol mountant and the cover slips. Slides of samples were stored in a dry oven at 45°C to 60°C for at least one week. Samples were observed under Leica ICC50 HD, which was connected to the camera and the image was directly taken and stored using Leica LAS EZ Software. The anatomical characters including root length, root diameter, cortex length, and vasculat bundle length were viewed and measured using this software.

RESULTS AND DISCUSSION

According to this study, planting medium type had a significant effect on the quantity of newly emerged roots. However, planting media had no significant effect on the number of primary, secondary, and tertiary roots. There were no significant differences in terms of cutting type among all parameters evaluated.

Number of newly emerged root

According to the results, the performance of newly emerged roots was considerably higher in a 50:50 mixture of peat and perlite (M3) (87 newly emerged roots) than in a 75:25 mixture of topsoil and sand (M1) with 34 newly emerged roots (Table 2). Luis (2014) stated that the length of the most significant root was measured in latosol, which may be related to the nutrient availability compared to sand. C2 (hardwood cutting) produced a greater quantity of newly emerged roots (81) when compared to semi-hardwood cutting (C1) which only produced 16 newly emerged roots (Table 3). Spethman (2007) reported that propagating roses and many tree species through long cuttings increased plant survival and development. Additionally, older wood provides better rooting, most likely due to the preformed root-initials (Osman Ibrahim, 2007). For treatment combinations, a mixture of 50% peat, 50% perlite, and hardwood cutting (M3C2) resulted in a substantially greater number of newly emerged roots (102) than a mixture of 75% topsoil, 25% sand, and semi-hardwood cutting (M1C1) (Table 4). According to Larsen (1997), the soil medium stimulates plant growth by providing physical support, moisture, and aeration. According to Okanlawon et al. (2016), increase in performance and growth of *Mussaenda philippica* may be attributed to nutrient availability within the soil combination of topsoil, sawdust, and poultry manure.

Table 2: Mean value of root growth parameters as affected by different planting medium

Planting Medium	Number of newly emerged root (NNER)	Number of primary root (NPR)	Number of secondary root (NSR)	Number of tertiary root (NTR)
M1 (75% topsoil : 25% sand)	33.6[b]	36.6[a]	215.5[a]	64.5[a]
M2 (33.33% topsoil : 33.33% sawdust : 33.33% sand	45.1[ab]	34.9[a]	237.5[a]	88.9[a]
M3 (50% peat : 50% perlite)	87.4[a]	42.2[a]	239.4[a]	171.3[a]

Note: Means followed by same letter within a column are not significantly different at p = 0.05 by Duncan's New Multiple Range Test (DNMRT) test

Table 3: Mean value of root growth parameters as affected by type of cutting

Type of cutting	Number of newly emerged root (NNER)	Number of primary root (NPR)	Number of secondary root (NSR)	Number of tertiary root (NTR)
C1 (Semi-hardwood)	16.4[a]	37.0[a]	252.7[a]	102.6[a]
C2 (Hardwood)	80.6[a]	38.6[a]	216.7[a]	107.3[a]

Note: Means followed by same letter within a column are not significantly different at p = 0.05 by Duncan's New Multiple Range Test (DNMRT) test

Number of primary root

The mixture of 50% peat and 50% perlite (M3) resulted in the highest primary roots (42). In contrast, the mixture of 33.33% topsoil, 33.33% sawdust and 33.33% sand (M2) yielded the fewest primary roots (35) (Table 2). According to Hassanein (2013), increased nutrition and moisture content result in increased rooting capacity, explaining stem cutting survival and rooting. M3, which is composed of peat and perlite, demonstrated the most outstanding performance in terms of primary root number as a consequence of the physical properties of peat and perlite. Costa et al al. (2017) claimed that a combination of peat and perlite was often utilised as a propagation medium for pot roses. Based on Table 3, when it comes to cutting type, C2 (hardwood cutting) produced the most primary roots (39) meanwhile, semi-hardwood cutting (C1) produced fewer primary roots (37). According to Jafar et al. (2008), no root was formed from softwood of *Capparis spinosa* var. *Parviflora* cuttings as compared to semi-hardwood cuttings. In terms of treatment combinations, a mixture of 50% peat and 50% perlite with semi-hardwood cutting (M3C1) gave the highest number of primary roots (44). On the other hand, a mixture of 33.33% topsoil, 33.33% sawdust, and 33.33% sand with semi-hardwood cutting (M2C1) resulted in the lowest number of primary roots (32) (Table 4). Kumar (2015) stated that the optimal propagation medium would supply sufficient nutrients and water, nourish the plant, and allow for gaseous exchange between the root atmosphere and the surrounding environment. Additionally, a combination of peat and perlite treatment resulted in a greater root length (Ali & Ahmad, 2017).

Table 4: Mean value of root growth parameters as affected by planting medium and cutting type

Planting Medium	Type of cutting	Number of newly emerged root (NNER)	Number of primary root (NPR)	Number of secondary root (NSR)	Number of tertiary root (NTR)
M1 (75% topsoil: 25% sand)	C1 (Semi-hardwood)	9.0[b]	35.5[a]	234.0[a]	45.0[a]
M1 (75% topsoil: 25% sand)	C2 (Hardwood)	50.0[ab]	37.3[a]	206.3[a]	80.2[a]
M2 (33.33% topsoil: 33.33% sawdust: 33.33% sand	C1 (Semi-hardwood)	20.2[ab]	32.3[a]	217.6[a]	114.8[a]
M2 (33.33% topsoil: 33.33% sawdust: 33.33% sand	C2 (Hardwood)	78.7[ab]	38.3[a]	257.4[a]	63.0[a]
M3 (50% peat: 50% perlite)	C1 (Semi-hardwood)	17.0[ab]	44.3[a]	323.7[a]	159.0[a]
M3 (50% peat : 50% perlite	C2 (Hardwood)	102.0[a]	40.0[a]	178.6[a]	178.6[a]

Note: Means followed by same letter within a column are not significantly different at p = 0.05 by Duncan's New Multiple Range Test (DNMRT) test

Number of secondary root

In terms of secondary root count, M3 has the most remarkable performance (239) and M1 has the lowest performance (216) (Table 2). Sydnor and Redente (2002) found that supplementing topsoil with organic wastes influences trace element absorption. According to van Duivenbooden et al. (1999), poor root growth resulted from increased penetration resistance in the topsoil. Thus, the lowest result for M1 may be attributed to the increase soil compaction in comparison to M3. In terms of cutting type, C1 produced the most secondary roots (253) while C2 has the least secondary roots (217) (Table 3). The most effective result in terms of days to sprouting was observed in semi-hardwood guava cuttings treated with IBA (Wahab et al., 2001). According to Hartmann et al. (2011), macadamia may be grown in the current season utilising leaf and semi-hardwood cuttings of mature wood. Additionally, semi-hardwood cuttings (partially mature stems) and hardwood cuttings (dormant mature stems) were utilised as cutting material from a healthy mother plant (Okunlola & Akinpetide, 2016). M3C1 has the most significant number of secondary roots (324), while M3C2 has the lowest number of secondary roots (179) (Table 4). The rooting media used may have an effect on the success of rooting and the quality of the resulting root system (Larsen, 1997). Compared to perlite and a 1:1:1 mixture of the three media, peat moss and sand media demonstrated substantially greater survival, rooting, and growth (Hassanein, 2013).

Number of tertiary root

M3 resulted in the most tertiary roots (171), whereas M1 has the fewest (65) (Table 2). Peat is defined by a high organic matter and carbon content (50% of its weight) (CSPMA, 2014). Thus, the usage of peat as planting medium mixture in M3 increase the number of tertiary root. In terms of cutting type, C2 generated the most tertiary roots (107), whereas C1 produced the fewest (103) (Table 3). Hartmann et al. (2011) reported that two to three-year-old Quince hardwood cuttings readily developed roots. Additionally, *F. benjamina* hardwood cuttings rooted more deeply than semi-hardwood cuttings (Okunlola & Akinpetide, 2016). M3C2 with contain 50% of peat generated the most tertiary roots (179) in comparison to M1C1, which produced the fewest tertiary roots (45) (Table 4). Peat is a type of soilless medium in its decomposed form. It is generally lightweight and has a high water retention capacity (Robbins & Evans, 2011). Increased water retention has a significant effect on plant growth due to the fact that water serves as a transport channel for nutritional components throughout the plant.

Root anatomical structure of *Ficus carica* cv. BTM6

Table 5 summarises the values for root length, root diameter, cortex length, and vascular bundle length. According to this study, semi-hardwood cutting with 75% topsoil and 25% sand resulted in the most extended root length (1.31 µm), whereas hardwood cutting with a mixture of 33.33 per cent topsoil, 33.33 per cent sawdust, and 33.33 per cent sand resulted in the shortest root length (0.74 µm). Additionally, semi-hardwood cuttings planted in a 50:50 mixture of peat and perlite resulted in the largest root diameter (1.51 µm). Meanwhile, a mixture of 50% peat and 50% perlite, semi-hardwood cutting generated the longest cortex (0.42 µm), whereas hardwood cutting produced the shortest (0.25 µm). With a mixture of 33.33 per cent topsoil, 33.33 per cent sawdust, and 33.33 per cent sand, semi-hardwood cutting generated the most extended vascular bundle length (0.60 µm). In contrast, hardwood cutting produced the shortest (0.34 µm). According to Lux et al. (2004), a bigger cortex increases water and solute movement within the root, resulting in more significant root growth. Additionally, primary root growth is more remarkable in semi-hardwood cuttings than in hardwood cuttings. Peat as a component of the planting media has an effect on the root performance of *Ficus carica* cv. BTM6 for both types of cuttings. It was discovered in this study that both types of cuttings had a more excellent value when planted in a mixture of 50% peat and 50% perlite. This may be because peat has a more significant proportion of organic materials. Additionally, peat has a high nutritional capacity and a high Cation Exchange Capacity (CEC) (Ranneklev & Baath, 2003). In addition, it has a high porosity of about 74%, which allows for significant water absorption (Bigelow et al., 2004). Moreover, since peat moss has a low bulk density (0.06 to 0.1 mgm⁻³), it is appropriate for usage in containers with a lower capacity. The increased root mass enabled the cuttings to absorb more nutrients and generate more leaves (Okunlola, 2013). Based on this study, semi-hardwood cutting achieved the highest values for all parameters compared to hardwood cutting. This is because hardwood cutting takes longer to develop roots than semi-hardwood cutting. Additionally, Danthu et al. (2002) found that the rooting capacity of the cuttings changed according to the propagation medium utilised.

Table 5: Root anatomical structure of *Ficus carica* cv. BTM6

Planting Medium	Type of cutting	Root length (μm)	Root diameter (μm)	Cortex length (μm)	Vascular bundle length (μm)
M1 (75% topsoil : 25% sand)	C1 (Semi-hardwood)	1.31	1.11	0.40	0.38
M1 (75% topsoil : 25% sand)	C2 (Hardwood)	0.86	0.20	0.27	0.38
M2 (33.33% topsoil : 33.33% sawdust : 33.33% sand	C1 (Semi-hardwood)	0.92	0.20	0.30	0.60
M2 (33.33% topsoil : 33.33% sawdust : 33.33% sand	C2 (Hardwood)	0.74	1.13	0.26	0.34
M3 (50% peat : 50% perlite)	C1 (Semi-hardwood)	1.24	1.51	0.42	0.40
M3 (50% peat : 50% perlite	C2 (Hardwood)	1.01	1.01	0.25	0.43

CONCLUSION

This study established that the planting media and cutting type have an effect on the root growth of *Ficus carica* cv. BTM6. For both types of cutting, a 50:50 mixture of peat and perlite promotes the most significant number of roots, root length, root diameter, and cortex length. Thus, a 50:50 mixture of peat and perlite was proposed for propagating *Ficus carica* using hardwood and semi-hardwood cuttings. For the type of stem cutting, semi-hardwood cutting was recommended to use as planting material.

ACKNOWLEDGEMENT

The authors would like to express their thanks and gratitude to the science officers from the Department of Plant Science, Kulliyyah of Science, International Islamic University Malaysia (IIUM), for their assistance and support throughout the study. This research was funded by the Research Initiative Grant Scheme (RIGS), IIUM (RIGS No. 15-128-0128).

REFERENCES

Acosta, M., Rocío Oliveros-Valenzuela, M., Nicolás, C., & Sánchez-Bravo, J. (2009). Rooting of Carnation Cutting. *Plant Signaling & Behavior*, *4*(3), 234-236. https://doi.org/10.4161/psb.4.3.7912

Ali, R., & Ahmad, H. (2017). The Effect of Growth Medium and Nodes Number on the Acclimatization Success of Tissue Propagated Potatoes Seedlings *Solanum tuberasum*. *International Journal of Science and Research (IJSR)*, *6*(4), 1478–1482. https://doi.org/10.21275/ART20172621

Ali, Z. A., Mustafa, N. S., Abdel-Raouf, H. S., El-Shazly, S. M., & El-Berry, I. M. (2013). Characterization of Some Fig Cultivars by Anatomical Traits on Both Leaves and Stems. *World Applied Sciences Journal*, *24*(8), 1065–1071. https://doi.org/10.5829/idosi.wasj.2013.24.08.13259

Bigelow, C., Bowman, D., & Cassel, D. (2004). Physical properties of three sand size classes amended with inorganic materials or sphagnum peat moss for putting green rootzones. *Crop Sci.*, *44*, 900-907. https://doi.org/10.2135/cropsci2004.9000

California Rare Fruit Growers, I. (2016). FIG. 1–5.

Costa, J. M., Heuvelink, E., & Van de Pol, P. (2017). Propagation by Cuttings. *Elsevier*, (1). https://doi.org/10.1016/B978-0-12-809633-8.05091-3

CSPMA. (2014). Canadian Sphagnum Peat Moss Association (CSPMA) Industry Social Responsibility Report. In *Canadian Sphagnum Peat Moss Association*.

Danthu, P., Soloviev, P., & Sarr, A. (2002). Vegetative propagation of some West African *Ficus* species by cuttings. *Agroforestry Systems*, *55*, 57–63. https://doi.org/10.1023/A

Dubrovsky, J. G., & North, G. B. (2002). Cacti biology and uses. In Regents of the University of California (pp. 41–56).

Euzébio de Souza, M., Leonel, S., Carvalho da Silva, A., Pacheco de Souza, A., Lopes Martin, R., Aki Tanaka, A., & de Souza, M. E. (2015). Carbohydrates, Growth and Production of " Roxo de Valinhos " Fig Tree in Initial Development under Irrigation Management. *American Journal of Plant Sciences*, *6*(6), 1126–1137. https://doi.org/10.4236/ajps.2015.68117.

Flaishman, M. A., Rodov, V., & Stover, E. (2008). The Fig: Botany, Horticulture, and Breeding. *Horticultural Reviews*, *34*, 113–196. https://doi.org/10.1002/9780470380147.ch2

Hartmann, H.T., D.E. Kester, F.T. Davies, & R.L. Geneve. (2011). Plant Propagation: Principles and Practices. 8th ed. New York: Prentice Hall.

Hassanein, A. M. (2013). Factors Influencing Plant Propagation Efficiency Via Stem Cuttings. *Journal of Horticultural Science & Ornamental Plants*, *5*(3), 171–176. https://doi.org/10.5829/idosi.jhsop.2013.5.3.1125

Jafar, M., Mehdi, B., Gask, R., & Shekafandeh, A. (2008). Influence of IBA and Cutting Length on Rooting Rate of Wild Caper (*Capparis spinosa* var . Parviflora) Stem Cuttings. *Middle Eastern and Russian Journal of Plant Science and Biotechnology*, *2*(2), 78–79.

Kumar, V. (2015). Growing media for healthy seedling production. *Van Sangyan*, *2*(9), 19-28. 10.13140/RG.2.1.4384.1764

Larsen, F. E. (1997). Propagating Deciduous and Evergreen Shrubs, Trees and Vines with Stem Cuttings. Cooperative Extension Washington State University. Retrieved from http://cru.cahe.wsu.edu/CEPublications/pnw0152/pnw0152.html

Lawrence, K. (2009). Basic Fig Tree Fact Sheet – Understanding Figs and How to Care for Them in Kansas. In Byron Wiley. Retrieved from bwiley@sbcglobal.net

Luis, A. (2014). Rooting of Hardwood Cuttings of Roxo de Valinhos Fig (*Ficus carica* L.) with Different Propagation Strategies. *Rev. Ceres, Viçosa*, *61*(6), 989-996. https://doi.org/10.1590/0034-737X201461060015

Mawa, S., Husain, K., & Jantan, I. (2013). Ficus *carica L. (Moraceae): phytochemistry, traditional uses and biological activities. Evidence-Based Complementary and Alternative Medicine*, 2013. https://doi.org/10.1155/2013/974256

Okanlawon, S. O., Babatunde, K. M., Salau, M. A., Adekanmbi, O. A., & Jmoh, A. R. (2016). Effects of Different Growth Media on Propagation of Horticultural Plant, Mussaenda philippica (Queen of Philippines). *International Journal of Current Research in Biosciences and Plant Biology*, *3*(7), 4–10. http://dx.doi.org/10.20546/ijcrbp.2016.307.002

Okunlola A I. (2013). The Effects of Cutting Types and Length on Rooting of *Duranta repens* in the Nursery. Global Journal of Human Social Science Geography*, Geo-Sciences, Environmental & Disaster Management*, *13*(3): 1–4

Okunlola, A. I., & Akinpetide, E. O. (2016). Propagation of *Ficus Benjamina* and *Bougainvillea Spectabilis* Using Different Media. *Advances in Agriculture and Agricultural Sciences, 2*(2), 21–27.

Osman Ibrahim, N. Y. (2007). Propagation of *Ficus nitida* L. by cuttings technique. Master thesis, University of Zalingei.

Paez-Garcia, A., Motes, C., Scheible, W.-R., Chen, R., Blancaflor, E., & Monteros, M. (2015). Root Traits and Phenotyping Strategies for Plant Improvement. *Plants, 4*(2), 334–355. https://doi.org/10.3390/plants4020334

Preti, F., & Giadrossich, F. (2009). Root Reinforcement and Slope Bioengineering Stabilization by Spanish Broom (*Spartium junceum* L.). *Hydrology and Earth System Sciences, 13*, 1713–1726. https://doi.org/10.5194/hess-13-1713-2009

Ranneklev, S.B., & Baath, E. (2003). Use of Phospholipid Fatty Acid to Detect Previous Self- heating Events in Stored Peat Moss. *Appl. Environ. Microbio, 69*, 3532-3539. https://doi.org/10.1128/AEM.69.6.3532-3539.2003

Relf, D., & Ball, E. (2009). Propagation by Cuttings, Layering and Division. Virginia Cooperative Extension.

Robbins, J. A., & Evans, M. R. (2011). Greenhouse and Nursery Series Growing Media for Container Production in a Greenhouse or Nursery Part I – Components and Mixes. *University of Arkansas Corperative Extension Service*, 1–4.

Seago Jr, J. L., & Fernando, D. D. (2013). Anatomical Aspects of Angiosperm Root Evolution. *Annals of Botany, 112*, 223–238. https://doi.org/10.1093/aob/mcs266

Spethmann, W. (2007). Increase of Rooting Success and Further Shoot Growth by Long Cuttings of Woody Plants. *Propag. Ornam. Plants, 7*(3), 160-166.

Sydnor, M.E.W., and Redente, E.F. 2002. Reclamation of High-elevation, Acidic Mine Waste with Organic Amendments and Topsoil. *Environ. Qual., 31*, 1528-1537. https://www.semanticscholar.org/paper/Reclamation-of-high-elevation%2C-acidic-mine-waste-Sydnor

van Duivenbooden, N., Pala, M., Studer, C., & Bielders, C. L. (1999). Efficient Soil Water Use: The Key to Sustainable Crop Production in Dry Areas of West Asia, and North Africa and Sub-Saharan Africa. Proceedings of the 1998 (Niger) and 1999 (Jordan) Workshops of the Optimizing Soil Water Use (OSWU) Consortium., 1–489

Wahab, F., Nabi, G., Ali, N., & Shah, M. (2001). Rooting Response of Semi-hardwood Cuttings of Guava (*Psidium guajava* L.) to Various Concentrations of Different Auxins. *Journal of Biological Sciences, 1*(4), 184–187. https://doi.org/10.3923/jbs.2001.184.187

Root Anatomy of Two *Ficus carica* Varieties as Affected by Different Planting Medium

Nurul-Aini, C.A.C.[1], Rozilawati, S.[1*] & Nik Nadira, N.N.R.[1]

[1]*Department of Plant Science, Kulliyyah of Science, International Islamic University Malaysia, Jalan Sultan Ahmad Shah, Bandar Indera Mahkota, 25200, Kuantan, Pahang, Malaysia.*
Corresponding author: firdawila@iium.edu.my

ABSTRACT

Ficus L. is classified under the genus in Moraceae family which is known as one of the largest genera of angiosperms with more than 800 species of trees, shrubs, or climbers in the tropics and subtropics worldwide. Since *Ficus* is known due to its edible fruits, they are acknowledged as the potential crop that can be commercialized especially in Malaysia. *Ficus carica* L. or Fig are referred to as one of the introduced crops in Malaysia. However, the information on its adaptability and agronomic practices is still lacking. Thus, this study aims to identify and describe the root anatomy of two *F. carica* varieties as affected by different planting mediums. Two selected *F. carica* varieties were used in this study which is *F. carica* cv. BTM6 and *F. carica* cv. TGF. This experiment was done at Glasshouse Nursery Complex (GNC), IIUM, Kuantan. Four different combinations of peat, topsoil, and sand were used as planting mediums. The root anatomical structure was prepared by using the basic histology technique. The characters were observed under the light microscope and analyze using LAS EZ software. The findings showed the planting medium gives the effect on primary, secondary, and tertiary root anatomical characters. The presence of peat moss promotes the highest in length and diameter of primary, secondary, and tertiary roots. Based on this study, the usage of 25 % sand: 25 % topsoil: 25 % peat moss: 25 % coco peat (M1), and 25 % sand: 50 % peat moss: 25 % topsoil (M2) was suggested to give the enhance of the performance for the root development of species studied.

Keywords: Root Anatomy, *Ficus Carica*

INTRODUCTION

Moraceae is classified as one of the largest genera in the mulberry family. Barolo et al. (2014) described the family Moraceae or an edible angiosperm plant family containing a milky latex. Common figs are classified under the genus *Ficus* and belong to the family Moraceae with over 1400 species classified into about 40 genera (Andersen & Crocker, 2016; Flaishman et. al., 2008; Stover et al., 2007; USDA, 2016). The major *Ficus* genus widely distributed in the Asiatic mainland (170 spp.), New Guinea (132 spp.), and Borneo (129 spp.). *Ficus* species habit can be described as deciduous and evergreen trees, shrubs, herbs, climbers, and creepers, and the life form was free standby tree, epiphytes, semi-epiphytes in the crevices, rheophytes, and lithophytes (Chaudhary et al., 2012). *F. carica* were first recorded as cultivated plants in southern Arabia in 2900 BC and mainly grown in the Mediterranean countries (Stover et al., 2007). Currently, the common *F. carica* become one of the most important crops and still grows wild in the Mediterranean basin (Mawa et al., 2013). According to Condit (1938), *F. carica* can be classified into four different groups that were Caprifig, Smyrna, White San Pedro, and common fig. In Malaysia, the common fig was widely planted. After all, it can adapt to this country's climate and one of the immemorial fruits among humans because it is edible (Aksoy, 1998).

The characteristics of *F. carica* are deciduous with the broad tree, ovate, three to five-lobed leaves (Mc Govern, 2002). On the other hand, the leaves are stipulated and petiolated with obovate, nearly orbiculate or ovate leaf blade, palmately lobed, cordate base, undulate or irregularly dentate margin, apex with acute to obtuse shape, and scabrous- pubescent surfaces. Fig produce seed-bearing fruit for functioning females, whereas others are functioning male and produce only pollen and pollen-carrying wasp progeny (Kjellberg et al., 1987; Janzen, 1979). The fig fruit (synocium or fig) and the reproductive system of *Ficus* species are different from other species.

Mawa et al. (2013) mentioned that the demand for Fig fruit is increasing due to the consumer interest in fruit because of the micronutrient content such as anthocyanin, pigment, flavonoid, and other phenolic compounds. *F. carica* cv. BTM6 and *F. carica* cv. TGF is classified as a famous variety compared to other varieties introduced in Malaysia due to its high survival and easy production of yields. Fig is also widely known and became one of the most important crops in Malaysia due to the high demand for seedlings and their fruit price. In order to obtain the high quality and yield of fig fruit, observation on the growth factor such as planting medium seems to be an important task. Therefore, the aim of this study is to describe the effect of different planting mediums on the root anatomical structure of two *F. carica* varieties.

MATERIALS AND METHODS

Experimental Site, Treatment and Layout

An experiment was conducted at Glass House and Nursery Complex, International Islamic University Malaysia from January 2017 until October 2017. The 2 months old seedling of *F. carica* cv. BTM6 (V1) and *F.carica* cv. TGF (V2) was used as planting material. Planting media used were combination of 25% peat moss: 25% coco peat: 25% sand: 25% topsoil (M1), 25% sand: 50% peat moss: 25% topsoil (M2), 25% sand: 50% coco peat: 25% topsoil (M3), 247 100% topsoil (M4). The plant was grown in polybag with size 20 x 20-inches. The experiment was outlined in a split-plot design with 5 replications where the type of growing medium and *F.carica* varieties was assigned as main plot and subplot, respectively. Plants were watered twice a day and organic fertilizer (15 grams/polybag) was applied every two weeks.

Primary and Secondary Roots (Rotary Microtome)

The primary, secondary and tertiary roots were prepared for the root anatomy process. Firstly, root samples were fixed into the AA solution right after harvesting for at least a week. The AA solution stands for 70% ethanol and glacial acetic acid with 1:3 ratios. Then, the middle part of the root (3 cm from tip) was cut using a small knife about 1 cm into the cross- section. Then, the entire sample was placed in the jar. The sample with AA solution in the jar was placed in a refrigerator that has a lower temperature than 4°C. After fixation, the specimen must undergo a dehydration process due to the hydrophobic paraffin wax. This process must be done slowly so that the delicate tissue was remaining firm. Therefore, the gradual dehydration in a series of alcohol or water solution of increasing strength must be prepared.

Embedding was one of the processes that insert the wax into the specimen. Firstly, put the specimen was arranged in a transversely position. Ensure the wax was completely hardening before undergo sectioning. Accelerated the wax hardening process by cooling it at a cold plate. After that, stored the embedding specimen at 40°C until sectioning. The rotary microtome was set up 8μm to10 μm in thickness. The blade angle can be increased if your observation was unclear.

The specimen that was already cut was floated in a warm water bath. It was float until the ribbons expand and lay completely flat. The water bath should be set at 36 to 42°C. It should be filled about 3 over 4 full. Progressive staining is used. Progressive staining was processed the tissue had been left in the stained until the desired depth of color is obtained.

Tertiary Roots (Sliding Microtome)

For the tertiary roots, the anatomical analysis was done using a sliding microtome. It was due to the cell size that was very small and thin and the structure was not rigid when using a rotary microtome. The steps of sliding microtome involve fixation, sectioning, clearing, staining, and dehydration process.

The sample was cut in a cross-section with a suitable size (<5mm). The desired specimen was left in the fixative at least for 48 hours. A fixative was used AA solution (glacial acetic acid and 70% ethanol) with a ratio of 1:3. Soak the sample into the AA solution. The fixative sample must be wash with running tap water for at least 20 minutes to clean up all the AA fixative solution. After cutting the tissue into the desired position, it must be pressed with cork before sectioning with a sliding microtome. The tertiary root tissue was put between the two corks. The cork was curved according to the shape of the tissue specimen. The position of the tissue was adjusted to avoid it from becoming unclear and tilted. Once a tissue starts to appear, support the end of the cork with a hair paintbrush and put the sample into the distilled water. Repeat this step to get the desired tissue. The thickness was about 25 µm – 35 µm according to the suitable thickness of the sample.

Then, samples were bleached into 5 % sodium hypochlorite (Clorox) solution for 10 minutes. The remaining clorox was washed up and soak sample in distilled water 3 times. It is important to wash up to avoid the staining process become unclear. After washed with clorox three times, the sample was soaked into the methylene blue for 5 minutes. After that, the sample was washed up with the distilled water to remove excess color. Next, samples were soaked into the safranin for 5 minutes. The sample was washed up with distilled water to remove any excess color. Next, the dehydration process was done using a series of alcohol. The sample was dehydrated by soaking for 5 minutes in a series of 50%, 70%, 95%, and 100%. The slice sample was mooved into the slide by using a paintbrush. The sample was mounted with Canada balsam onto glass slides and coverslips. The sample was arranged in an oven-dry at 55°C for at least two weeks.

Observation under Light Microscope
The observation of the specimen was done under a microscope Leica ICC50 HD that connected with a CCTV camera and the images were taken and saved by using Leica LAS EZ software. The analysis of anatomy character was measured by using this software.

RESULT AND DISCUSSION
The root anatomical character of different *F. carica* cv. varieties was studied in different types of planting medium. The combination of planting medium and varieties showed significant difference for all the parameters taken except vessel diameter.

Anatomy of Primary Roots
Based on the microscopic view (Fig 1), the combination of 25% coco peat: 25% peat moss: 25% sand: 25% topsoil with *F. carica* cv. BTM6 (M1V1) shows the highest value in root length (3.93 µm) and root diameter (3.28 µm). The highest value in root diameter and root length showed that the combination of 25% coco peat: 25% peat moss: 25% sand: 25% topsoil with *F. carica* cv. BTM6 (M1V1) had a higher function in structural support and transportation of water. The cell diameter and length of the root had correlated with the plant's vegetative growth that increases in the lower part of the plant enhance the growth of upper parts of the plant. It was supported by a study from Mayaki et al. (1976) that showed there is a relation between the plant height and the root diameter of the plant, especially in soybean. While another research that has been conducted by Samejima et al. (2005) reported that the root biological yield of the tropical rice was closely related to the above-ground part of the plant. For instance, a large volume of roots together with the strong water absorption from the plant increase in productivity of the tropical rice itself (Samejima et al., 2004).

However, the combination of 25% topsoil: 50% coco peat: 25% sand with *F. carica* cv. TGF (M3V2) shows the lowest in root length (2.48 µm) and diameter (2.05 µm) indicates that this planting medium shows the lowest structural support and water transportation of the plants. It was due to the very high-water holding capacity which causes poor aeration in the root zone. It will later affect the oxygen of the root (Ilahi & Ahmad, 2017). Coco peat also can easily affect air density and water retention (Abad et al., 2002). The increase in the diameter of roots also can be related to the nutrient uptake by the plant. The nutrient uptakes from the planting medium were important for plant growth and productivity. Lopez- Bucio et al. (2003) explained that the bioavailability of nutrient uptake by the plant in the soil solution may determine the root growth, root proliferation, and specific functional responses that depend on the nutritional status of the plant.

The combination of 25% coco peat: 25% peat moss: 25% sand: 25% topsoil with *F. carica* cv. BTM6 (M1V1) also recorded the highest value in cortex and xylem length with 1.81 μm and 1.8 μm, respectively. Based on this study, the cortex and xylem increase as the increase in root length and diameter. Meanwhile, the combination of 25% sand: 50% coco peat: 25% topsoil with *F. carica* cv. TGF (M3V2) resulted lowest in cortex length (1.07 μm), root diameter (1.1 μm), and also root length (1.43 μm). According to Lux et al. (2004), the larger size of the cortex increases the radial movement of water and solutes in the root. Myburg et al. (2001) described the xylem as one of the vascular bundle tissue that mainly located in the stems and roots of moderns plants whereas the xylem is the most important function of transporting water and minerals nutrients. Thus, 25% coco peat: 25% peat moss: 25% sand: 25% topsoil with *F. carica* cv. BTM6 (M1V1) has the highest water transportation compare to other treatments due to the highest in xylem length. The usage of planting medium with peat moss (M1 and M2) increases in vessel diameter compared to the planting medium with the absence of peat moss (M3 and M4). It was supported by Hacke et al. (2005) that increase in vessel frequency, increase in the transportation and storage of water in the plant.

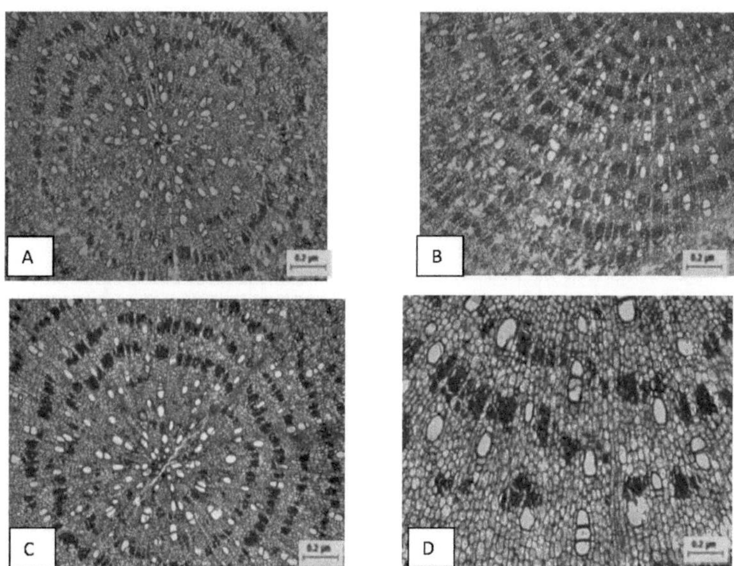

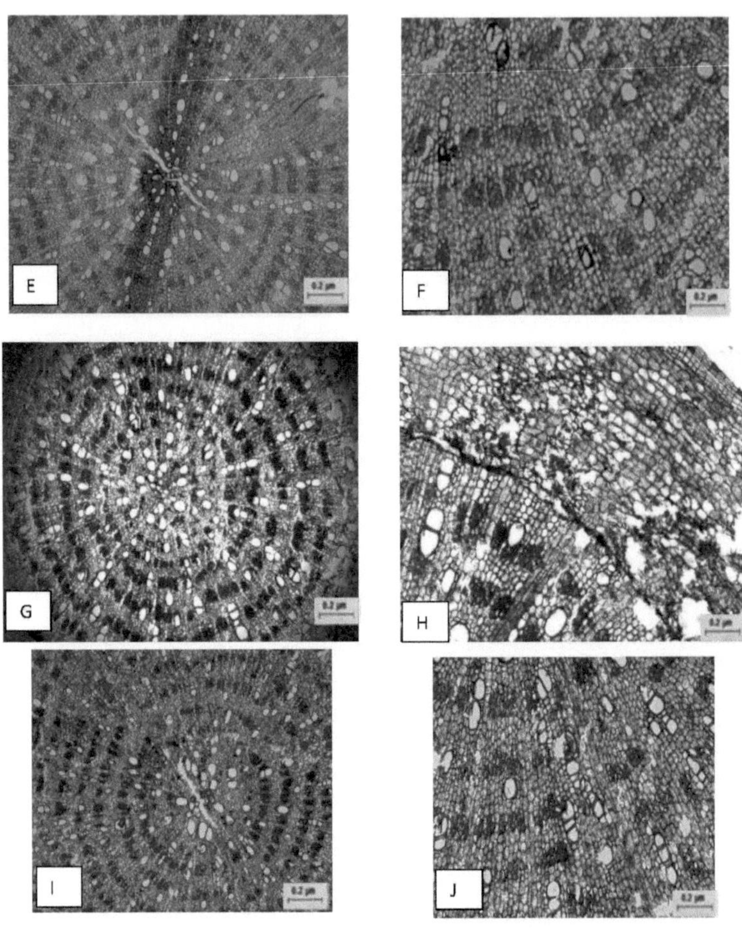

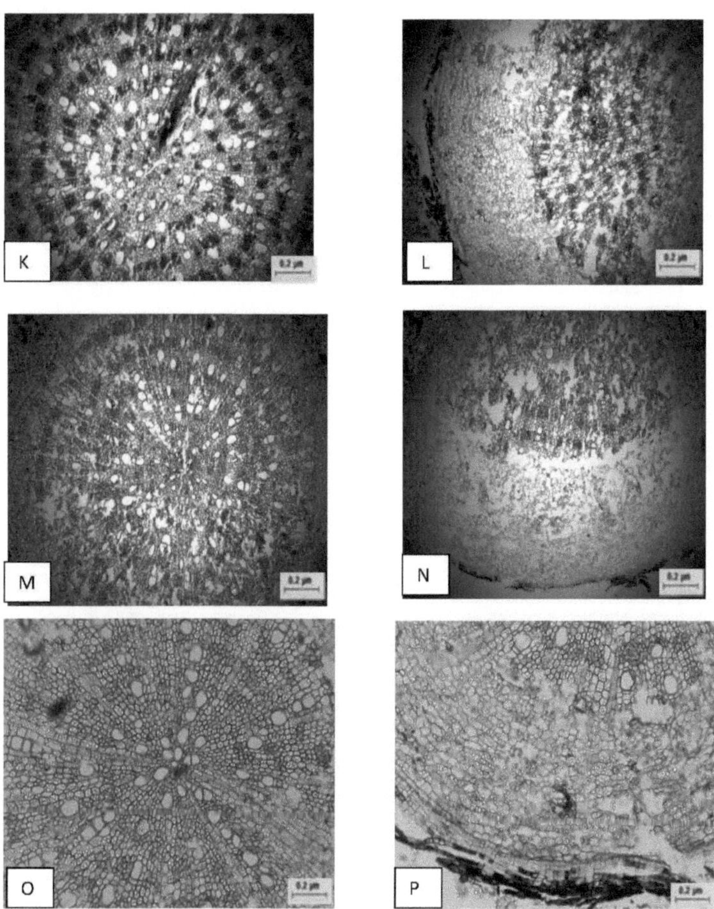

Fig. 1: Primary Root Anatomical Structure (Cross Section of Middle and Side Part, A&B) M1V1 C&D)M1V2 E&F)M2V1, G&H)M2V2, I&J)M3V1, K&L)M3V2, M&N)M4V1, O&P)M4V2 (4x Magnification) (scale bar at 0.2μm).

Anatomy of Secondary Roots

The function of the secondary root almost the same as the primary root which is very important for anchorage and structural support of the plant (Myburg & Sederoff, 2001). Going through the data, it is obvious that there was a positive effect of using peat moss as a planting medium compares to other treatments. Based on this study, the combination of 25% coco peat: 25% peat moss: 25% sand: 25% topsoil

(M1) gives the highest in root length (2.47 µm) and root diameter (2.40 µm) (Fig 2). The root structural and functional properties can be conveyed based on the size of the root. The cell size also can be the indicator for the axial growth (Cahn et al., 1989), growth duration (Eisssenstat et al., 2000), root anatomy (Jordan et al., 1993), and water transport activity (Vercambre et al., 2002).

Combination of 25% coco peat: 25% peat moss: 25% sand: 25% topsoil (M1) gives the highest performance of cortex length and xylem length with 1.13 µm and 1.44 µm, respectively. The cortex length was important for the absorption of nutrients in the plants. Commonly, the superficial root that larger in cortex length are capable to provide nutrients to the plant during the dry period (Gourlat & Marcati, 2008). This study also showed that the media combination of 25% coco peat: 25% peat moss: 25% sand: 25% topsoil (M1) and a combination of 25% sand: 50% peat moss: 25% topsoil (M2) can enhance the water and absorption of roots due to the increase of the vessel diameter and frequency. It was supported by a study from Leonell & Tecchio (2009) that added organic matter probably improved the development of fig tree root system contribute to the high in yields production.

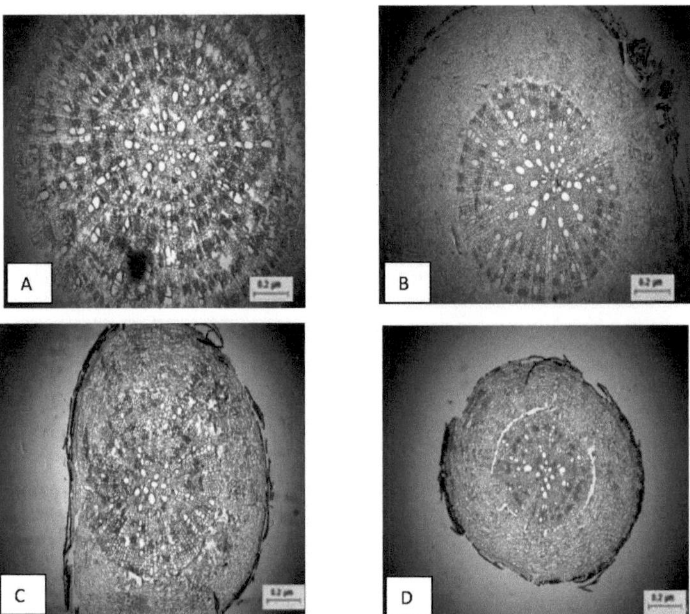

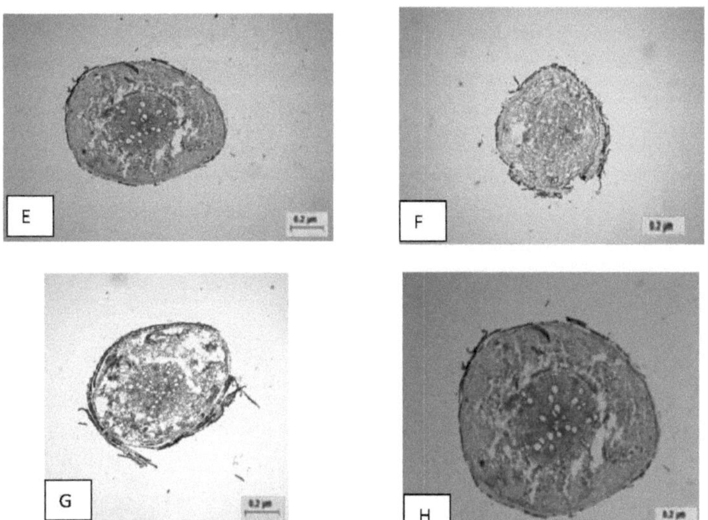

Fig. 2: Secondary Root Anatomical Structure (Cross Section of Middle and Side Part, A&B) M1V1 B) M1V2 C) M2V1, D) M2V2, E) M3V1, F) M3V2, G) M4V1, H) M4V2) (4 Magnification) (scale bar at 0.2μm).

Anatomy of Tertiary Roots

Based on the morphological observations, tertiary roots were the smallest root compared to primary and secondary roots. Due to this circumstance, the anatomical character for axial parenchyma and rays was unclear compare with the other primary and secondary roots. These anatomical characters might not fully develop because tertiary roots were still in the primary growth of the cell.

Based on this study, the combination of 25% sand: 50% coco peat: 25% topsoil (M3) gives the highest value of root length (1.12 μm) and root diameter (0.97 μm) (Fig 3). The largest number of roots hair constituent of the good root system for plant growth (Mc Cully,1999). For the cortex length and xylem length, the combination of 25% sand: 50% coco peat: 25% topsoil (M3) shows the highest value with 0.68 μm and 0.56 μm, respectively. The cell size and xylem length correlated with each other as the cell length and diameter increase also increase the xylem length (Twumasi et al., 2009). According to Chimungu et al. (2014), the large cortical cell size would enhance the root's drought tolerance. It was supported from the study by Prince et al. (2017), the increase in metaxylem vessels would enhance the efficiency of water uptake in soybean.

The vessel that is longer and bigger in diameter reduces the risk of embolism during drought, or during and after frost (Comstock & Sperry, 2000; Tyree & Zimmermann, 2002). The vessel diameter and frequency of root planted in 100 % topsoil (M4) resulted in the highest compared to other treatments with 0.04 μm and 14, respectively. It proved that M4 had the highest water uptake in tertiary roots due to increased vessel number. Other than that, tertiary roots were also important to enhance the uptake of nutrients and water by increasing the absorptive surface area. The size of the root, whether longer and wider in size, usually resulted in greater water transport capacity, which could reduce the water stress in the upper parts of plants during times of evaporative demands. Changes in water aeration of plants stressed by high aeration may increase the water uptake by tertiary roots.

This change considered that one way for the plant to defend itself in tolerate stress is by the increase in ability to absorb water (Abdul Qados, 2010). Generally, the tertiary root planted in a combination of 25% sand: 50% coco peat: 25% topsoil (M3) may induce by the high-water absorption which may improve the number of the root. The decrease in the number of tertiary roots in a combination of planting medium with peat moss (M1 and M2) was due to the high-water holding capacity that avoids the stress of tertiary plant roots.

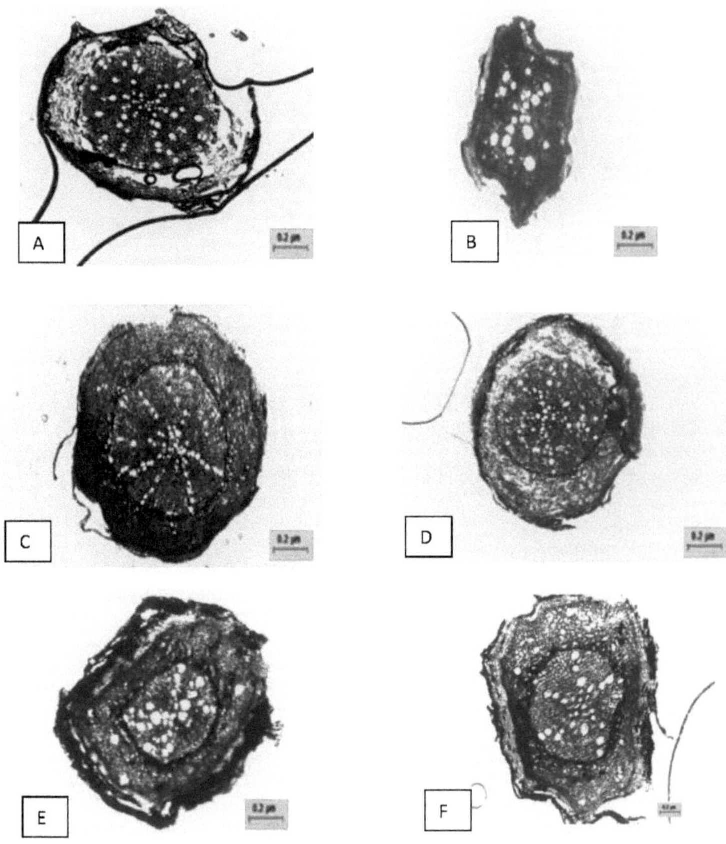

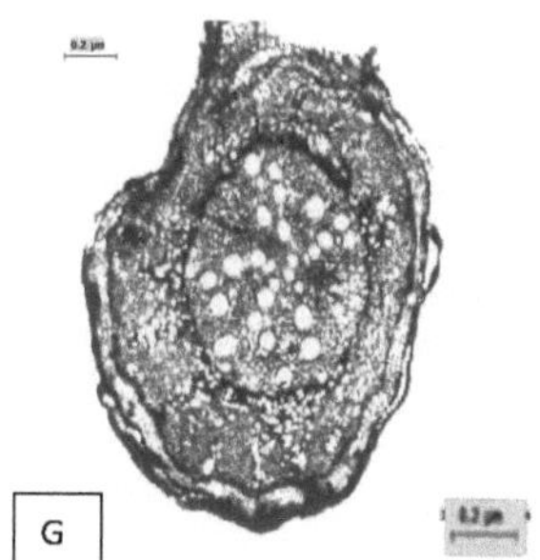

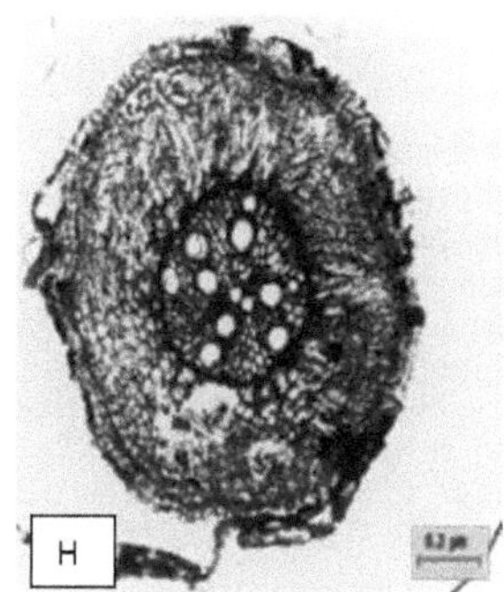

Fig. 3: Tertiary Root Anatomical Structure (Cross Section of Middle and Side Part, A&B) M1V1 B) M1V2 C) M2V1, D) M2V2, E) M3V1, F) M3V2, G) M4V1, H) M4V2) (10x Magnification) (scale bar at 0.2μm).

CONCLUSION

The planting medium gives the effect on primary, secondary, and tertiary root anatomical character. The presence of peat moss promotes the highest in length and diameter of primary, secondary, and tertiary roots. Based on this study, the usage of 25 % sand: 25 % topsoil: 25 % peat moss: 25 % coco peat (M1), and 25 % sand: 50 % peat moss: 25 % topsoil (M2) was suggested to give the enhance the performance of root development of *F. carica*.

REFERENCES

Abad, M., Noguera, P., Puchades, R., Maquieira, A. & Noguera, V. (2002). Physico-chemical and chemical properties of some coconut coir dusts for use as a peat substitute for containerised ornamental plants. *Bioresource Technology.* 82(3), 241-245.

Abdul Qados, A.M.S (2010). Effect of salt stress on plant growth and metabolism of bean plant *Vicia faba* (L.). *Journal of the Saudi Society of Agricultural Sciences.* 10, 7-15.

Aksoy, U. (1998). Why fig? An old taste and a new perspective. *Acta Horticulturae*, 480, 25-35.

Anderson, P.C. & Crocker, T.E. (2016). The fig. *Institute of Food and Agricultural Sciences.* Retrieved from http://edis.ifas.ufl.edu/mg214.

Barolo, M.I., Mostacero, N.R. &Lopez, S.N. (2014). *Ficus carica* L. (Moraceae): An ancient source of food and health. *Food Chemistry*, 164, 119-127.

Cahn, M.D., Zobel, R.W. & Bouldin, D.R. (1989). Relationship between root elongation rate and diameter and duration of growth of lateral roots of maize. *Plant and Soil*, 119, 271–279.

Chaudhary, L.B. Sushakar, J.V., Kumar, A. Bajpai, O., Tiwari, R. & Murthy, G.V.S. (2012). Synopsis of the genus *Ficus* L. (Moraceae) in India. *Taiwania*, 57(2), 193-216.

Chimungu J.G., Brown, K.M. & Lynch, P.J. (2014). Large root cortical cell size improves drought tolerance in maize. *American Society of Plant Biologist.* 166(4), 2166–2178.

Comstock, J. P. & Sperry, J. S. (2000). Theoretical considerations of optimal conduit length for water transport in vascular plants. *New Phytologist,* 148, 195–218.

Condit, I. J. (1938). Parthenocarpy in the fig. Proc. Amer. Soc. Hon Set, 36, 401.

Eissenstat, D.M., Wells, C.E., Yanai, R.D. & Whitbeck, J.L. (2000). Building roots in a changing environment: implications for root longevity. *New Phytologist*, 147, 33–42.

Flaishman, M. A., Rodov, V., & Stover, E. (2008). The fig: botany, horticulture and breeding. *Horticultural Reviews*, 34.

Gourlat, S.L. & Marcati, C.R. (2008). Anatomia comparada do lenho em raiz e caule de Lippia salviifolia Cham. (Verbenaceae). *Revista Brasileira de Botanica*, 31, 263- 275.

Hacke, U.G., Sperry J.S. & Pittermann J. (2005). Efficiency versus safety trade offs for water conduction in angiosperm vessels versus gymnosperm tracheids. In: N.M. Holbrook & M.A. Zwienniiecki (eds.) *vascular transport in plants*. Elsevier Incorporated, Amsterdam, 333-354.

Ilahi, W.F.F. & Ahmad, D. (2017). A study on the physical and hydraulic characteristics of cocopeat perlite mixture as a growing media in containerized plant production. *Sains Malaysiana*, 46(6), 975–980.

Janzen D.H. (1979). How to be a Fig. *Annual Review of Ecolology and Systematics*, 10, 13–51.

Jordan, M.O., Harada, J., Bruchou, C. & Yamazaki, K. (1993). Maize nodal root ramification: absence of dormant primordia, root classification using histological parameters and consequences on sap conduction. *Plant and Soil*, 153, 125–143.

Judd, L.A., Jackson, B.E. & Fonteno, W.C. (2015). Advancement in root growth measurement technologies and observation capabilities for container- Grown plants. *Plants*, 4, 369-392.

Kjellberg F., Gouyon P.H., Ibrahim M., Raymond M. & Valdeyron G. (1987). The stability of the symbiosis between dioecious figs and their pollinators: a study of *Ficus carica* L. and *Blastophaga psenes* L.," *International Journal of Organic Evolution*, 41(4), 693–704.

Leonell, S. & Tecchio, M.A. (2009). Cattle manure fertilization increase fig yield. *Sci. Agric. (Piracicaba, Braz.)*, 6 (6), 806-811.

Lopez-Bucio J., Cruz-Ramirez A. & Herrera-Estrella L. (2003). The role of nutrient availability in regulating root architecture. Curr. Opin. *Plant Biol*, 6, 280–287.

Lux, A., Luxová, M., Abe, J., & Morita, S. (2004). Root cortex: structural and functional variability and responses to environmental stress. *Root Research*, 13(3), 117–131.

Mawa, S., Husain, K. & Jantan, I. (2013). *Ficus carica* L. (Moraceae): *Phytochemistry, traditional uses and biological activities*. Hindawi Publishing Corporation. Kuala Lumpur: Malaysia.

Mayaki, W.C., Teare, I.D. & Stone, L.R. (1976). Top and root growth of irrigated and nonirrigated soybeans. *Crop Science*, 16, 92–94.

McCully M.E. (1999). Roots in soil: unearthing the complexities of roots and their Rhizospheres Annu. Rev. Plant Phys. *Plant Mol. Biol.*, 50, 695–718.

McGovern, T.W. (2002). The fig – *Ficus carica* L. *Cutis*, 69 (339), 40.

Myburg, A.A. & Sederoff, R.R. (2001). *Xylem structure and function*. North Carolina, USA: Nature Publishing Group.

Prince S.J., Murphy, M., Mutava R.N., Durnell L.A.,Valliyodan, B., Shannon, J.G. & Nguyen H.T. (2016). Root xylem plasticity to improve water use and yield in water-stressed soybean. *Journal of Experimental Botany*.

Samejima, H., Kondo. M., Ito, O., Nozoe, T. & Osaki, M. (2004). Root-shoot interaction as a limiting factor of biomass productivity in new tropical rice lines. *Soil Science and Plant Nutrition*, 50, 545-554.

Samejima, H., Kondo, M., Ito, O., Nozoe, T. & Osaki, M. (2005). Characterization of root systems with respect to morphological traits and nitrogen-absorbing ability in the new plant type of tropical rice lines. *Journal of Plant Nutrition*, 28, 835-850.

Stover, E., Aradhya, M. Ferguson, L. & Crwasosto, C.H. (2007). The fig: Overview of an ancient fruit. *Hort Science*, 42(5), 1083-1087.

Twumasi, P. Schel, J.H.N & Ieperen, W.V. (2009). Differential effects of temperature on stem length and xylem vessel length in Zinnia elegans. *Journal of Horticultural Science & Biotechnology*, 84 (5), 531- 535.

Tyree, M. T. & Zimmermann, M. H. (2002). *Xylem Structure and the Ascent of Sap*. 2nd Edition, Springer, Berlin, Germany. 250.

USDA. (2016). *Ficus carica* L.: Classification. Retrieved from http://plants.usda.gov/core/profile?symbol=FICA.

Vercambre G, Doussan C, Pagès L, Habib R & Pierret A. 2002. Influence of xylem development on axial hydraulic conductance within Prunus root systems. *Trees: Structure and Function*, 16, 479–487.

Nutrient Requirement and Fertilizer Management for Fig (*Ficus carica* L.)

Muhamad Fahmi Yunus, Mohd Razik Midin and Md Hoirul Azri Ponari*

[1]Department of Plant Science, Kulliyyah of Science, International Islamic University Malaysia, Jalan Sultan Ahmad Shah, Bandar Indera Mahkota, 25200, Kuantan, Pahang, Malaysia.
**Corresponding author: mhazri@iium.edu.my*

ABSTRACT

Ficus carica L. (common fig) is a crop species that has been cultivated in many countries especially in warm and dry climates. However, it is considered a new species cultivated in Malaysia for commercial value. Thus, it is crucial to understand the nutritional demand of the fig plant so that the best agronomic strategies such as the right nutrient sources, appropriate nutrient requirement doses, and the best time of nutrient application can be determined for high yield production. Fertilizers need to be adequately supplied to the soil to replenish the nutrients removed from the harvested yield and prevent nutrients deficiency symptoms that lead to a lower yield and quality of fruit. Nutrient deprivation will hinder plant metabolism hence, decreasing the vegetative growth, mineral concentration, and chlorophyll content in leaves. Overall, the percentage of all elements during fruit development is higher compared to mature leaves. Nitrogen is the largest amount of essential element required by fig plants, followed by potassium, phosphorus, calcium, and magnesium respectively. Nitrogen is the largest amount of minerals in the plant tissue and contributes to the dried weight of the plant. The percentage of nitrogen is high during the fruit and leaves development because it is required for cell division, metabolism of plants, and inorganic feed increases both fruit quantity and consistency. Phosphorus involves in many primary plant functions such as flower initiation and fruit production. This chapter will discuss the nutrient requirement and management as well as the deficiency symptoms and effects of nutrient deficiency in *F. carica*.

Keywords: Plant nutrition, Essential nutrients, Deficiency symptoms

INTRODUCTION

F. carica, also known as the common fig, is the only plant aside from *F. sycomorus* in the genus Moraceae that is cultivated for its pear-shaped edible fruits. The fig fruits are really hollow fleshy receptacles containing hundreds of male and female flowers, or botanically known as a syconium. Fig plants are divided into four types which are Common type, San Pedro type, Smyrna type and Caprifig type. Common type *F. carica*, or fig is widely cultivated in many countries for its high nutritional and medicinal values. Fig has been reported to contain a good source of dietary fibre and minerals, amino acids, antioxidants, vitamins, and valuable phytochemicals such as phenolics, anthocyanins and carotenoids. Plus, it has a high level of calcium, ferum and copper, with fat and cholesterol-free (Jagtap and Bapat 2019; Shatnawi et al., 2019; Crisosto et al. 2011). Fig has been used in treatment for several diseases namely cough, skin diseases, cancer and it is also used as a painkiller (Pavía et al. 2020).

The cultivation of figs is important because this fruit is consumed both dry and fresh. Turkey, Egypt, Algeria, Morocco, and Iran are the countries with the highest production of figs because the fig trees are abundantly cultivated throughout this region and are well adapted to drought and high temperatures conditions (FAOSTAT, 2013). The great economic importance of this fruit in countries such as Turkey and Egypt have promoted the research and development of agronomic practices for fig cultivation (de Souza, 2003). Thus, this review will attempt to shed more light on the nutrient requirement, fertilizer management for figs as well as the visual symptoms of nutritional deficiencies of the fig tree.

MORPHOLOGY OF FIG

Morphologically, fig plant is soft woody shrubs or small trees with smooth and grey-colored bark (Crisosto et al. 2011; Barolo, Mostacero, and López 2014). According to Crisosto *et al.*, (2011), the fig leaves were distinguished traits among fig varieties, arranged in alternate and have seven lobes with a hairy surface. The foliage is palmate, large and petiolate with three to seven lobes (Crisosto et al. 2011). It has a distinct feature which is a milky white latex secretion that contains ficin, a type of digesting enzyme (Baraket et al. 2009). In addition, the fig is a gynodioecious species which means they are having hermaphrodite flowers on the same species and others containing only pistillate flowers (Crisosto et al. 2011). Before bearing fruits, the globular-shaped structure of syconium consists of hundreds of little unisexual flowers (Beck 2008). The drupelets fruit of fig undergoes development from ovaries in a closed syconium (Al-Khayri et. al., 2018). Fig is identified by the fruits which contain infertile seeds that do not undergo pollination. Figs have different types of cultivars which bear breba crop (first crop) and second crop which is called the main crop (Al-Khayri et. al., 2018).

NUTRITION REQUIREMENT AND MANAGEMENT

Proper management of soil nutrients is crucial to ensure the strong growth of plants. It will affect the end product of the plants such as lower yield and reduced yield quality. Plant nutrients can be divided into two categories: macronutrients and micronutrients. Primary macronutrients consist of nitrogen (N), phosphorus (P), potassium (K) and these elements are needed in large amounts by plants as they become limiting factors in plant development. On the other hand, secondary macronutrients are nutrients required in a lesser amount than those three NPK elements (Lai et al. 2019). Their presence is important since an insufficient supply of any micronutrients will lead the plant to exhibit a deficiency symptom.

Table 1 is a summary of the data taken from Brown (1994), on the study of the nutrient concentration in *F. carica* through different developmental stages. The data in the table shows the mean nutrient concentrations in fig leaves from healthy sites for 3 years. For macronutrients, Calcium (Ca) and N content were the highest in fig leaves and fruit followed by K, Magnesium (Mg) and P. This can indicate that Ca and N are required in a large amount for healthy leaves and fruit development of *F. carica* as compared to other macronutrients. In plant physiology, Ca is involving in plant structural functions such as in the cell wall and membranes. Apart from that, Ca also serves as a counter-cation for inorganic and organic anions in the vacuole and act as an intracellular messenger in the cytosol (White and Broadley, 2003).

For micronutrients, Iron (Fe) content was the highest, followed by Manganese (Mn), Zink (Zn) and Copper (Cu). In terms of developmental stages, the N, P, K elements are the highest during flowering and fruit development. This suggests that applying NPK fertilizer especially with a higher ratio of N and K during these two stages is important to ensure the plants can fruit properly and have enough nutrients for survival. For micronutrients, there is not much variability during different developmental stages. It might be because of the small amount needed for the development process. However, the high levels of Mn in the data might indicate toxicity.

Mendoza-Castillo et al. (2019), repeating a similar study and the results were identical to Brown's finding (Table 2). In the absence of nitrogen, the stem elongation and diameter increase in size with only small changes of value. The same condition applies to the plant height, internode length, as well as dry weight of root, stem, and leaves. This is due to the fact that N is the important component of plants such as proteins, coenzymes, nucleic acid, chlorophyll and most secondary metabolites. N has the highest concentration in plant tissues which is around 5% total dry weight of a plant (Garza-Alonso et. al., 2019). This is also proven in the research in which a high amount of nitrogen fertilizer increases fruit production. On the other hand, Moreno et. al., (1998) reported nutrients requirement results for mature leaves (Table 2). Micronutrients for fruit development and maturity are needed in a small amount. The concentration requirement for Iron (Fe) is 287 mg Kg^{-1}, Manganese (Mn) 290 mg Kg^{-1}, Zinc 90 mg Kg^{-1} and Copper (Cu) is 11 mg Kg^{-1} (Mendoza-Castillo et. al., 2019).

Table 1: Nutrient concentration in *F. carica* through different developmental stages (Brown, 1994).

Type of nutrient	Nutrient element	Developmental stage			
		Flowering	Fruit development	Fruit maturity	Post-harvest
Macronutrient (%)	N	2.3	2.1	1.6	1.5
	P	0.14	0.12	0.11	0.09
	K	1.4	1.0	0.7	0.7
	Ca	3.0	3.0	2.9	3.5
	Mg	0.7	0.7	0.8	0.8
Micronutrient (ppm)	Zn	12	12	12	9
	Cu	6	6	5	4
	Mn	80	90	145	150
	Fe	120	110	125	78

Table 2: Macronutrients requirement for fruit development and maturity and mature leaves.

Macronutrients	Amount of element (%)	
	Fruit development and maturity (Mendoza-Castillo et al. 2019)	Mature leaves (Moreno et. al., 1998)
Nitrogen (N)	3.40	2.64 - 2.66
Phosphorus (P)	0.58	0.14 - 0.15
Potassium (K)	1.10	0.11 - 0.12
Calcium (Ca)	3.40	0.34 - 0.35
Magnesium (Mg)	0.64	0.04 - 0.05

NUTRITION DEFICIENCY SYMPTOMS OF FIG

N, P, K and Mg are mobile nutrients. These nutrients can be translocated from older tissues to newly generated tissue. Thus, the deficiency symptoms of these nutrients would appear first in older leaves of the lower part of the stem. Mg and N are the basic constituent of chlorophyll molecules which involve in the photosynthesis process. Mn involves in the water-splitting process afters Photosystem II (PSII) fixes light to initiate the conversion of CO_2 and water into carbohydrates in the photosynthesis process. Cu is essential for photosynthesis and mitochondrial respiration. While boron is associates with the flowering and fruiting process in plants. Zn is required for energy production and activation of enzymes for the synthesis of protein. It also plays a role in biomembrane structural integrity (Mendoza-Castillo et. al., 2019; Brown, 1994).

Table 3: The essential plant nutrients requirement, its function, deficiency symptoms for fig tree cultivation.

Nutrient	Function	Plant deficiency symptoms	Effect of nutrient deficiency	Reference
Nitrogen (N)	Basic elements for protein and chlorophyll synthesis which plays a major role in photosynthesis.	For mature leaves, chlorosis appeared on the edge of the left central lobe.	Trees have stunted growth and smaller size. Reduced number of branches with smaller size.	(Manoel et al. 2018; Motwani et al. 2015; Garza-Alonso et al. 2019; Pavía et al. 2020)

		Lack of leaves and sparse foliage. Accelerating the senescence process. detachment appeared in the mature leaves of the lower part of the stem.		
Phosphorus (P)	P generates energy, sugars, and nucleic acids. Associate with cellular metabolism and fruit development. The basic constituent of chlorophyll.	The symptom can be seen on the older leaves since P is mobile in plants. The intense green colouration on mature leaves, chlorosis without a defined pattern on mature leaves, leads to necrosis on mature leaves. An inadequate level of P causes stress to the plants. As a result, leaves turn purple because of the surge of anthocyanin.	Premature leaf drops, delay the ripening of fruits, smaller stem and stunted growth.	(Garza-Alonso et al. 2019; Pavía et al. 2020)
Potassium (K)	Regulate plant turgidity and osmotic pressure for stems. Basic constituent of cytoplasm.	For mature leaves, chlorosis appeared in the margins and spread to the lobe's ribs. Necrosis. Curly margins inwards for mature leaves. Green band presence at mature leaves from the base of the stem.	No fruits formation.	(Manoel et al. 2018; Garza-Alonso et al. 2019; Pavía et al. 2020)

Calcium (Ca)	Required for cell wall and membrane development. Cofactor of enzymes, associate with the ATP and phospholipids hydrolysis process.	Leaf lobes on young leaves showed chlorosis and deformity.	Smaller stem.	(Garza-Alonso et al. 2019)
Magnesium (Mg)	Chlorophyll production.	Interventional chlorosis started at the central part of the lateral lobes. Occurred at third and fourth mature leaves from the lower part of the stem.	Slow plant growth.	(Manoel et al. 2018; Mendoza-Castillo et al. 2019; Garza-Alonso et al. 2019)

Fig. 1: Deficiency symptoms of macronutrients in fig tree leaves. (a) **Control;** (b) **N deficiency;** (c) **P deficiency;** (d) **K deficiency;** (e) **Ca deficiency;** (f) **Mg deficiency (Garza-Alonso et al. 2019)**

A study using the missing element technique in a controlled hydroponic system found that the fig plants showed specific visual symptoms in the absence of different macronutrients. The vegetative growth of the fig tree was affected most in the absence of N, followed by P, K, and Ca. Inadequate availability of N, K, and Mg for fig tree nutrient uptake decreased the chlorophyll content in the lower stratum of plants. In addition, the absence of some elements leads to a greater accumulation of other ions in the different tissues of the plant such as the leaf, stem, and root due to interactions of antagonism and synergism of the elements involved (Garza-Alonso et al. in 2019)

APPLICATION OF BENEFICIAL MICROORGANISMS ON FIG PLANT

In the soil system, there is an area called the rhizosphere. Mineral uptake is a process involving a translocation pathway where mineral elements pass up the xylem until they reach the area of their sinks, such as adolescent leaves, developing flowers, fruits and seeds, apical and lateral meristems, and individual cells for storage. Minerals are unloaded at fine vein terminuses through diffusion. They are picked up by cells through active uptake. Rhizosphere also contains root exudates which cause soil microorganism especially rhizobacteria to colonize the roots. Rhizobacteria are considered as one of the important communities in rhizosphere microbiota. Some of these rhizobacteria were found able to influence plant development as well to protect the plant roots against phytopathogens. These plant-beneficial rhizobacteria are known as plant growth-promoting rhizobacteria (PGPR). PGPR can influence plant growth through a variety of direct and indirect mechanisms. For instance, PGPR stimulates plant growth directly by enhancing nutrient availability for the plant nutrient uptake and producing numerous growth regulators that can modulate the plant's hormonal level. Protection of plants from phytopathogens, improving soil structure, and protecting plants from the inhibitory effects of various abiotic stress such as bioremediating the heavy metal polluted soils and degrading toxic xenobiotic compounds, are examples of how PGPR can indirectly enhance plant growth. (Vejan et. al., 2016; Tariq et. al., 2017; Kumari et. al., 2019). Thus, the role of PGPR in improving crop production like figs has become more important nowadays.

A few PGPR activities could directly increase the nutrient availability in soil. PGPR can fix nitrogen independently or symbiotically and enhances plant nitrogen uptake derived from Biological Nitrogen Fixation (BNF). An example of a free-living nitrogen-fixing organism is *Azospirillum*, often associated with crop yields improvement, including Ficus tree (Osman et. al., 2010). Some PGPR can assist in the enhancement of phosphate acquisition by plants. These PGPR increase availability of phosphate ions in the soil, which can be easily taken up by the Ficus plants. The principal mechanism of this process is the production of organic acids by the PGPR that can solubilize the inorganic phosphorous in the soil. The inoculation of phosphate solubilizing bacteria effectively increases phosphate uptake by the Ficus plant. PGPR Strains such as *Bacillus, Pseudomonas,* and *Rhizobium* are reported among the most efficient phosphate solubilizers (Vejan et. al., 2016).

Phytohormones produce by PGPR such as indole-3-acetic acid (IAA) can alter root architecture and promote plant development. This exogenous IAA mainly helps in the enhancement of root length, increase in the number of root branches, root hairs and root laterals that aid in the acquisition of nutrients from the soil. Besides that, an increase in plant root volume will directly increase the amount of root exudation that provides additional nutrients to support the growth of rhizosphere bacteria. (Glick, 2012). IAA also plays many important roles from early plant growth, such as embryo development, leaf formation, and root initiation, to fruit development and leaf abscission. It has been found out that a vast species of rhizobacteria is capable of synthesizing exogenous IAA and affects physiological changes in plants, especially on roots. The root of the Ficus plant also was reported inhabited by beneficial microorganisms that can synthesize the IAA and promote seed germination and growth (Rodge et. al., 2016). A previous study demonstrated that a consortium of PGPR composes of Bacillus polymyxa, Rhizobium leguminosarum, and Psedmonase flurscence has the potential to promote root formation in Ficus species cuttings. The stimulation of root growth by these PGPR can be correlated to the production of indole-3-acetic acid (IAA) by the bacteria (Hend et. al., 2016).

In terms of symbiosis relationship, the fig provides food for microorganisms while these beneficial microorganisms help the uptake of water, minerals and protect the plant from phytopathogen. There are hundreds of known arbuscular mycorrhizae that already undergo symbiosis with plants thousands of years ago. Among the common species of arbuscular mycorrhizae, *Glomus caledonium, Glomus margarita, Glomus intraradices* and *Glomus clarium* are effective for plant growth (Comlekcioglu et al. 2008).

The symbiotic relationship between arbuscular mycorrhizae with vascular plants able to improve plant health and tolerance to abiotic stresses such as drought. Besides that, it also helps in the uptake of elements especially macronutrient and micronutrients from the soils. According to Comlekcioglu *et al.,* (2008) mycorrhizal inoculation on the fig plants caused increases in shoot length, the dried weight of root, zinc, and phosphorus absorption. Therefore, the inoculation of mycorrhizal species has a positive impact on the growth of fig plants by helping the nutrient uptake in the soil.

As a result, the colonization of microbes and plants builds a symbiosis relationship. The study by Comlekcioglu *et al.,* (2008), found that arbuscular mycorrhiza fungi helps to boost nutrient uptake for fig trees via root infection. Comlekcioglu *et al.,* (2008) noticed that *G. clarium* and *G. margarita* have better phosphorus absorption while *G. caledonium* improved the uptake of Zn. The threads of mycorrhiza form a network around the young root, or some of them will penetrate the root cells.

On top of that, other than cooperating fertilizers to supply adequate essential nutrients to fig, applying plant growth promoting rhizobacteria (PGPR), as well as mycorrhizae can also contribute to the well growth of the plants. These tiny living organisms can be cooperated into the soil by using manure or biofertilizer as it contains these microorganisms as the main ingredients (Abd-Alhamid et al. 2015). Research performed by (Comlekcioglu et al. 2008) revealed that inoculation of mycorrhizae positively contributes to maximizing nutrient absorption and plant development. Dry weights of fig "Alkuden" cultivar that were inoculated with various species of mycorrhizae had been obtained, it was found that the obtained weights for root and the shoot parts were higher compared to the control ones. Moreover, the concentration of several nutrient elements such as phosphorus, zinc, iron, and copper were found to be higher for figs that have involved in a symbiotic relationship with beneficial mycorrhizae.

Table 4 shows the effect of mycorrhizae on plant development. This data was taken from (Comlekcioglu et al. 2008) study where they treat *F. carica* in several species of arbuscular mycorrhiza fungi to see their effects on mineral uptake of fig. In terms of shoot and root dry weight, fig inoculated with mycorrhizae had a significant increase compared to the controlled one. The highest reading was 0.37 g which is from *G. margarita* species. For root infection, the highest mycorrhizal infection in root was recorded from *G. caledonium* species which is 86.7%. In terms of shoot and root height, the difference was not really big between controlled and inoculated fig. Lastly, in terms of nutrient concentration, the level of Fe, Zn and Cu increased significantly in inoculated fig compared to control.

Table 4: Mycorrhizal effect on plant development

Effect on plant development	Presence of mycorrhizae	
	Control	Mycorrhizal infection
Shoot and root dry weight	0.21 g	0.37 g
Root infection	0 %	86.7 %
Shoot and root height	No significant difference between both situations. However, it is significantly higher in some species of mycorrhizae.	
Concentration of nutrients	Increased significantly with mycorrhizal infection.	

Table 5 shows the effect of PGPR on plant development. El-Morsi and Abdel-Monaim, (2015) treated *F. carica* with several PGPR species for a period of 3 months. Inoculation using *Azotobacter* showed the disease severity towards *Fusarium oxysporum* decreased by more than 53% with the presence of PGPR. Other than that, the plant height, number of leaves per plant and dry weight also increased significantly in inoculated fig.

Table 5: Effect of PGPR on plant development

Component	Presence of PGPR	
	Control	*Azotobacter* inoculation
Disease severity	75.69 %	22.47 %
Plant height	10.08 cm	26.02 cm
Number of leaves per plant	6.00	14.33
Dry weight	23.14 gm · plant^{-1}	42.14 gm · plant^{-1}

CONCLUSION

Adequate and balanced nutrition is essential for the production of commercial fig fruit. The fig plants required a complete macronutrient and micronutrient for optimum growth and showed specific visual symptoms in the absence of each element. The concentration of nutrients needed may be influenced by several factors such as the origin of cultivars, fertilizer, and the composition of the soil. Encouragement should also be given to the application of organic fertilizer and plant beneficial microorganisms such as plant growth-promoting rhizobacteria and arbuscular mycorrhizal fungi in the fig agronomic practice due to the good impact on crop productivity and ecosystem.

REFERENCES

Abd-Alhamid, N., Hassan, H. S. A., Laila, F., Haggag, & Hassan, M. A. (2015). Effect of mineral and bio-fertilization on vegetative growth, leaf mineral contents and flowering of manzanillo olive trees. *International Journal of ChemTech Research*, 8(11), 51–61.

Al-Khayri, J. M., Jain, S. M., & Johnson, D. V. (2018). *Advances in Plant Breeding Strategies: Fruits: Volume 3*. Springer International Publishing.

Baraket, G., Saddoud, O., Chatti, K., Mars, M., Marrakchi, M., Trifi, M., & Salhi-Hannachi, A. (2009). Sequence analysis of the internal transcribed spacers (ITSs) region of the nuclear ribosomal DNA (nrDNA) in fig cultivars (*Ficus carica* L.). *Scientia Horticulturae*, 120(1), 34–40.

Barolo, M. I., Ruiz Mostacero, N., & López, S. N. (2014). Ficus carica L. (Moraceae): An ancient source of food and health. *Food Chemistry*, 164, 119–127.

Beck, H. (2008). Tropical Ecology. *Encyclopedia of Ecology*, 3616–3624.

Brown, P. H. (1994). Seasonal variations in fig (Ficus Carica L.) leaf nutrient concentrations. *Horticulture Science*, 29(8),871–73.

Comlekcioglu, S., Akpinar, C., Bayazit, A.B., Ortas, I., & Küden, A.B. (2008). Effect of mycorrhizae applications on the mineral uptake in 'Alkuden' (01-IN-06) fig genotype. *Acta Horticulturae*, (772), 513–520.

Crisosto, H., Ferguson, L., Bremer, V., Stover, E., & Colelli, G. (2011). Fig (Ficus Carica L.). Postharvest biology and technology of tropical and subtropical fruits: Cocona to Mango. Woodhead Publishing Limited.

de Souza, R. M. M. (2003). Fig culture techniques. *Acta Horticulturae*, 605:99–101.

El-Morsi, M. E. A, & Abdel-Monaim, M. F. (2015). Effect of bio-agents on pathogenic fungi associated with roots of some deciduous fruit transplants and growth parameters in New Valley Governorate, Egypt. *Journal of Plant Protection Research*, 55(2), 126–35.

FAOSTAT (United Nations Organization for Agriculture and Food, statistics) (2013). Based on FAO statistical data.http://faostat.fao.org/.

Garza-Alonso, C. A., Olivares-Sáenz, E., Gutiérrez-Díez, A., Vázquez-Alvarado, R. E., & López-Jiménez, A. (2019). Visual symptoms, vegetative growth, and mineral concentration in fig tree (Ficus carica L.) under macronutrient deficiencies. *Agronomy*, 9(12), 787.

Jagtap, U. B., & Bapat, V. A. (2019). Exploring phytochemicals of Ficus carica L. (Fig). *Reference Series in Phytochemistry*, 1–16.

Kumari B., Mallick M. A., Solanki M. K., Solanki A. C., Hora A., & Guo W. (2019). Plant Growth Promoting Rhizobacteria (PGPR): Modern Prospects for Sustainable Agriculture. In: Ansari R., Mahmood I. (eds) Plant Health Under Biotic Stress. Springer, Singapore.

Lai, C. H., Settinayake, A. R. H., Yeo, W. S., Lau, S. W., & Jong, T. K. (2019). Crop nutrients review and the impact of fertilizer on the plantation in Malaysia: A mini review. *Communications in Soil Science and Plant Analysis*, 50(17), 2089–2105.

Manoel, E. de, Sarita, L., Andrea, C. da, Adilson, P. de, Ana, H. M., Roberto, L. V., & Adriana, A. T. (2018). Nutrient absorption march and accumulation of nutrients in developing Roxo de Valinhos fig (Ficus carica L.) tree, cultivated under different water regimes. *African Journal of Agricultural Research*, 13(16), 834–843.

Mendoza-Castillo, V. M., Pineda-Pineda, J., Vargas-Canales, J. M., & Hernández-Arguello, E. (2019). Nutrition of fig (Ficus Carica L.) under hydroponics and greenhouse conditions. *Journal of Plant Nutrition,* 42(11–12): 1350–65.

Moreno, D. A., Pulgar, G., Víllora, G., & Romero, L. (1998). Nutritional diagnosis of fig tree leaves. *Journal of Plant Nutrition,* 21(12), 2579–88.

Motwani, H. V., De Rosa, M., Odell, L. R., Hallberg, A., & Larhed, M. (2015). Aspartic protease inhibitors containing tertiary alcohol transition-state mimics. *European Journal of Medicinal Chemistry*, 90, 462–490.

Osman, S. M., & El-rhman Abd, I. E. (2010). Effect of organic and bio N-fertilization on growth, productivity of fig tree (Ficus carica, L.). *Research Journal of Agriculture and Biological Sciences*, 6(3),319-328.

Pavía, Y. L. F., Patricia, S., Pavía, F., & Lúa, A. M. (2020). Nutrient deficiencies induced in fig tree Cv. Neza in hydroponic conditions. *Revista Mexicana Ciencias Agrícolas*, 11(3), 581–92.

Rodge, S. P., Sable, S. K., Salve, S. K., Sawant, S. A., & Patil, N. P. (2016). Isolation and characterization of PGPR from roots of Ficus religiosa growing on concrete walls and its effect on plant growth in drought condition. *International Journal of Current Microbiology and Applied Sciences*, 5(9), 583–593.

Shatnawi, M., Shibli, R. A., Shahrour, W. G., Al-Qudah, T. S., & Taleb, A. Z. (2019). Micropropagation and Conservation of Fig (Ficus Carica L.). *Journal of Advances in Agriculture*, 10, 1669–1679.

Tariq, M., Noman, M., & Ahmed, T. (2017). Antagonistic features displayed by plant growth-promoting rhizobacteria (PGPR): A review. *Genetics and Molecular Biology*, 35, 38–43.

Vejan, P., Abdullah, R., & Khadiran, T. 2016. Role of plant growth promoting rhizobacteria in agricultural sustainability - A review. *Molecules*, 21, 1–17.

White, P. J., & Broadley, M. R. (2003). Calcium in plants. *Annals of Botany*, 92,487–511

Bris Soil Improvement Using Organic Manure with Chemical Fertilizer for Growth and Yield Development of Fig (*Ficus carica* L.)

Fatin Munirah Azmi[1], Nur Shuhada Tajudin[1*], Rozilawati Shahari[1], Che Nurul Aini Che Amri[1], Mayzaitul Azwa Jamaludin[2]

[1]*Department of Plant Science, Kulliyyah of Science, International Islamic University Malaysia, Kuantan 25200, Pahang Malaysia*
[2]*Institute of Tropical Agriculture & Food Security, Universiti Putra Malaysia 43400 UPM Serdang, Selangor, Malaysia*
Corresponding author: nurshuhada@iium.edu.my

ABSTRACT

Proper use of manure and compost is essential for plant production and environmental perspective. Correct applications of this substances help to improve the chemical fertilizer efficiency and the soil physical properties, which later promote the growth of crops. The study was carried out to observe the effects of chicken manure on soil chemical and physical properties and the response of different application rates on the yield of fig trees planted on BRIS and mineral soil. The BRIS (Beach Ridges Interspersed with Swales) soil was amended with different rates of chicken manure; 10% (T2), 20% (T3), 30% (T4) and 50% (T5) and 0% (T1) acted as a control. The growth response of the fig trees was observed biweekly for nine months after transplant. The soil and leaf nutrient contents of amended and non-amended of both types of soil showed a significant difference at $p < 0.05$. Amending BRIS soil with 30% of chicken manure can been suggested to improve the soil health for fig planting with considering that it can be easily accessed and low in cost for agronomic practices.

Keywords: BRIS soil, Mineral soil, Fig, Nutrients, Growth response.

INTRODUCTION

Fig trees are cultivated in approximately 40 countries around the world, predominantly in Mediterranean region, for fruit consumption and production (McGovern, 2002). Fig (*Ficus carica* L.) is one of the five plants mentioned in the Quran, along with olives, grapes, pomegranate and dates. Fig is known as a nutritious fruit as it contains beneficial minerals and vitamins, traditionally used for its medicinal benefits as metabolic, respiratory, cardiovascular and anti-inflammatory remedy (Patil and Patil, 2011; Mawa et al. 2013). Recently in Malaysia, the fruit and leaves have been commercialized and introduced to the public as a nutritious fresh fruit and its leaves used as healthy tea. It has high economic value since the fruit produced and marketed are highly priced in Malaysia due to its limited productivity.

A good soil property is important in order to determine a success growth of crops. Sand, clay, silt, peat, chalk and loam are common types of soil based on the dominating size of the particles within a soil. Each soil plays a different role in physical and chemical characteristics and determines the health of the crops. BRIS (Beach Ridges Interspersed with Swales) is a sandy type of soil which mostly found near the coastal area in Pahang with area of 36,017.17 ha, Terengganu around 67,582.61 ha, and in Kelantan about 17,806.20 ha (Armanto et al., 2013). It is known as problematic soil with a weak structure, lack of nutrients, low water retention and high soil temperature (Basri et al., 2013; Ishaq et al., 2013; Nur Amirah et al., 2015). This makes the soil unfavourable for a crop to grow with. In Malaysia, BRIS soil is not utilized for crop production due to their inherent poor fertility. Until now, only tobacco and kenaf have been planted as profitable crops over the years (Roslan et al., 2011). Previously, studies related to BRIS soil improvement had been conducted on crops such as maize (Ishaq et al., 2014), kenaf (Malisa et al., 2011; Basri et al.,

2013) and sweet potato (Ishaq et al., 2014). In order to vary crops that can be planted on this type of soil, a study is needed to determine its requirement for a good growth of other crops.

Amending chicken manure can act as valuable sources of organic matter in soil, improving soil structure or tilth, increases water-holding capacity of coarse textured sandy soils, improves drainage in fine-textured clay soils, provides a source of slow release nutrients, and promotes growth of beneficial soil organisms (Okoli and Nweke, 2015; Ojeniyi et al., 2013). Chicken manure is preferred amongst other animal wastes because of its high content of nitrogen, phosphorus and potassium needed for crop growth (Duncan, 2005). Dikinya and Mufwanzala (2010) stated that chicken manure application enhanced soil fertility by way of increasing exchangeable bases in the soils. Study conducted by Awodun (2007), shows that poultry manure can effectively increase soil fertility, yield and nutrient content of fluted pumpkins planted in sandy clay loam soil. Therefore, chicken manure amendments can be suggested to improve the physico-chemical characteristic of the BRIS soil and the soil structure of mineral soil.

Currently, there is a lack of information regarding fig planting on different types of soil in Malaysia. Improving BRIS soil can be optimized with identification of sufficient rate of chicken manure for fig planting. Hence, this study was carried out to observe the effects of chicken manure on BRIS soil chemical and physical properties, leaf nutrient concentrations, growth, and yield of fig.

MATERIALS AND METHODS

Soil and experimental site preparation
Pot experiment was carried out in June 2016 until March 2017 at the Glasshouse and Nursery Complex, International Islamic University Malaysia (IIUM), Kuantan Campus, Pahang (3°51'08.3"N, 103°18'42.0"E). The BRIS soils used in this study were obtained from Rudua series (siliceous, isohyperthermic, grey, typic haplorthod) located at Cerok Paloh, Pahang (3°37'26.7"N 103°23'01.0"E). The soil collected were sieved to remove any remaining roots and large rocks. Composted dried chicken manure were selected for this study as a slow-release source of macro- and micronutrients and as a soil amendment (Aviafic Basal). Composting process initially "cools" the manure and litter material, meaning it reduces the ammonia content so it will no longer burn plants. It also reduces the total volume, weight and odor of the pile. Additionally, composting stabilizes nutrients enabling a slow, long-term release over a few years. The temperatures generated in the composting process also helped to kill most pathogens and weed seeds.

Experimental design
The experiment was conducted in June 2016. Five different rates of chicken manure (volume/volume) were used, which are 0% (T1) as control, 10% (T2), 20% (T3), 30% (T4) and 50% (T5). Fig saplings of Brown Turkey Modified (BTM6) cultivar were propagated through hardwood cuttings. The experiment was done in Randomized Complete Block Design (RCBD) with five replications for each treatment. All the polybags received 10g of 15:15:15 NPK fertilizer biweekly and irrigated with an automated dripping system twice per day.

Soil and chicken manure analysis
The soil samples were taken before planting (June 2016) and during harvesting (March 2017). The physical and chemical properties of soil samples from each treatment were analyzed. The soil samples were air-dried and sieved under 2mm sieve. Soil pH was determined using soil: water ratio of 1: 2.5 and pH meter (Metler Toledo FE20, Switzerland). Particle size analysis was analyzed using Particle Size Analyzer (PSA) (Mastersizer, 2000). Total C was determined using a carbon analyzer (LECO CR-412, USA and the total N was determined by using Kjedahl method (Bremner and Mulvaney, 1982). EC was determined using soil: water ratio of 1:5 (Rhoades, 1982) and EC meter (EZDO 6061, Taiwan) and CEC was determined using neutral ammonium acetate extraction methods (Schollenberger and Simon, 1945). Extractable P, K, Ca, Mg, Fe, Cu, Zn, and Mn were determined by Mehlich III method (Mehlich, 1984). N, P, and K were

measured by Auto Analyzer (AA) (Lachat QuikChem FIA 8000 series, USA). Ca, Mg, Fe, Cu, Zn and Mn were measured by using atomic absorption spectrometry (AAS) (Perkin Elmer 400, USA). Chemical analysis of chicken manure solely was done in the same way as the soil.=

Leaf nutrients analysis

After nine months of planting, the third leaf was sampled randomly from three branches of each tree. The samples were oven-dried, grind and sieved under 2mm sieve. The lamina of leaf samples was selected to perform the analysis of N, P, K, Ca, Mg, Fe, Cu, Zn and Mn using wet digestion method. N, P, and K were measured by auto analyser (AA) (Lachat QuikChem FIA 8000 series, USA). Ca, Mg, Fe, Cu, Zn and Mn were measured by using atomic absorption spectrometry (AAS) (Perkin Elmer 400, USA).

Growth response

For the growth performance observations, the height of the tree, stem girth, number of branches, number of leaves, number of fruits and survivability of each tree were taken biweekly until March 2017.

Statistical analysis

Data were subjected to analysis of variance (ANOVA), using Statistical Analysis System (SAS 9.2) to determine the effect of the treatments. The least significant difference (LSD) at $p < 0.05$ was used to compare the mean data of each parameter.

RESULTS AND DISCUSSION

Properties of chicken manure

Table 1 summarizes chemical composition and nutrient properties of the chicken manure used in this study. The pH of the chicken manure is 7.02 which is considered as neutral. It contained a high level of total N (1.4%) and total C (22.91%). Chicken manure generally contains the highest proportion of total N and is well supplied with micronutrients (Martin and MacRae, 2014; Duncan, 2005). The C/N ratio value was 16.37% that is within a desirable range (13.9–19.6%) for nutrient mineralization and promoting rapid composting (Kumar et al, 2010). The high EC (12.92 dS/m) in chicken manure is attributable to higher salt levels of N, K, Ca and Mg, which is proportionally high. The CEC were high (24.79 cmol₊/kg) due to a presence of relatively higher number of exchangeable cations in chicken manure used. Similar findings by Biratu et al. (2018) stated the chicken manure can contribute favorable qualities, including high levels of nutrients including total nitrogen (3.6%) and CEC (26 cmol₊/kg), cations such as Ca (9 cmol₊/kg) and K (2 cmol₊/kg), most micronutrients (Fe, Mn, Zn, Cu) and act as a potential soil amendment to address the low levels of nutrients and organic matter in the soils. High amounts of nutrients and carbon content in chicken manure used can compensate for the nutrient deficiency in BRIS soil.

Table 1: Chemical composition of chicken manure used.

Properties	Amount
pH	7.02
C/N	16.37
EC (dS/m)	12.92
CEC (cmol₊/kg)	24.79
Total C (%)	22.91
Total N (%)	1.4
Extractable P (%)	0.11
Extractable K (%)	1.54
Extractable Ca (%)	0.84
Extractable Mg (%)	0.6
Extractable Fe (ppm)	146.5
Extractable Cu (ppm)	37.3
Extractable Zn (ppm)	125.7
Extractable Mn (ppm)	220.1

Effect of amendments on soil physico-chemical characteristics

The addition of chicken manure to BRIS and mineral soil were observed to significantly increase the physico-chemical characteristics of soil ($p<0.05$), especially at the highest application rate (BT5) (Table 2). The soil texture for BRIS soil ranges from sand to loamy sand, silt loam to loam in mineral soil. The soil texture was improved in BRIS soil due to the mixing effect of organic amendments, which lowered the percentage of sand (Dikinya and Mufwanzala, 2010; Kidinda et al., 2015). Chicken manure added into the soil acts as cementing agent between soil particles, thus improving its aggregation and porosity. In loamy soil, this process of aggregation improves water infiltration, soil tilth, and subsequently plant growth (Mandal et al., 2013). The pH for BRIS soil that has zero chicken manure was 5.53, which indicates an acidic type of soil. Roslan et al. (2010) stated that BRIS soil is an acidic soil with pH values ranging from 4.8 to 5.6. The pH was observed to increase with increased application rate of manure for both types of soil. The addition of most types of organic waste to acidic soils is one of a reliable strategy to increase the soil pH (Carmo et al., 2015). This is due to the ion exchange reactions, which occur when terminal OH⁻ of Al or Fe^{2+} hydroxyl oxides are replaced by organic anions, which are decomposed products of the manure such as malate, citrate and tartrate (Eshoo and Bell, 1992; Pocknee and Summer, 1997; Hue and Amiens, 1989).

Angelova et al. (2013) stated that the direction of the change in soil pH as a result of treatment application reflected the initial pH of amendment material. The increase in pH could be due to the high pH value of chicken manure (Table 1) compared to non-amended soils. The highest pH recorded is 8.82±0.01 (BT5), which received 50% rate of chicken manure. With the addition of chicken manure, the number of exchangeable bases increased with the increasing application rate. The CEC of BRIS soil increased from 0.80cmol₊/kg in BT1 to 6.12cmol₊/kg in BT. This showed that application chicken manure increased the CEC of the BRIS soil and mineral soils significantly ($p<0.05$). This is due to the increase in the amount of organic matter addition, which leads to pH impairment and cations availability (Khairi et al., 2011). Similar studies were reported by Mandal et al. (2013), which stated CEC increased as BRIS soil being amended with composted poultry litter by improving its pH. This enable the soil to retain nutrients longer in the root zone and can be effectively used by rice plants. The higher the CEC, the better the soil ability to hold added cations, which ultimately improves the soils potential for fig growth.

In BT5, 50% application of chicken manure resulted in highest total N (0.24%) and extractable P (0.063%), K (0.248%), Ca (0.45%), Mg (0.19%) compared to BT1 (Table 2). Statistical analysis showed that the increase of P and N concentration due to chicken manure addition was significant ($p < 0.05$). This is in comparable with other studies (Dikinya and Mufwanzala, 2010; Adeniyan and Ojeniyi, 2005; Adenawoola and Adejoro, 2005; Davis et al., 2006), where P was significantly increased by chicken manure application. The total N content is higher than the P and this is due to nitrogenous compounds, for example ammonia found in the chicken manure, which is released during decomposition of the chicken manure, which is released during decomposition (Dikinya and Mufwanzala, 2010). Mineral soils have higher nutrients content compared to BRIS soil due to its ability to hold nutrients and prevent leaching. Nevertheless, low macronutrients of the BRIS soil (Basri et al., 2013) are caused by rapid leaching process of these nutrients (Ekhwan and Mazlin, 2009; Kadir et al., 2013).

Table 2: Physico-chemical properties of BRIS soil prior to planting.

Treatment (v/v)	(0%)	(10%)	(20%)	(30%)	(50%)
Properties	**BT1**	**BT2**	**BT3**	**BT4**	**BT5**
Soil Texture	Sand	Sand	Sand	Loamy Sand	Sand
pH	5.53±0.01[e]	6.99±0.01[d]	8.60±0.01[b]	8.58±0.01[c]	8.82±0.01[a]
C/N	9.50±0.01[a]	4.43±0.01[e]	5.60±0.01[d]	7.42±0.01[b]	6.38±0.01[c]
EC (dS/m)	0.15±0.01[e]	1.84±0.16[d]	4.41±0.62[b]	3.40±0.04[c]	6.77±0.66[a]
CEC (cmol₊/kg)	0.80±0.12[d]	1.81±0.16[cd]	2.69±0.54[c]	4.00±0.81[b]	6.12±0.61[a]
Total C (%)	0.19±0.01[e]	0.31±0.01[d]	0.84±0.01[c]	1.41±0.01[b]	1.53±0.01[a]
Total N (%)	0.02±0.01[e]	0.07±0.03[d]	0.15±0.02[c]	0.19±0.03[b]	0.24±0.01[a]
Extractable P (%)	0.001±0.001[e]	0.014±0.001[d]	0.043±0.007[c]	0.053±0.005[b]	0.063±0.007[a]
Extractable K (%)	0.011±0.009[d]	0.048±0.003[cd]	0.105±0.062[bc]	0.176±0.076[ab]	0.248±0.058[a]
Extractable Ca (%)	0.03±0.01[d]	0.15±0.02[c]	0.35±0.09[ab]	0.24±0.10[bc]	0.45±0.06[a]
Extractable Mg (%)	0.01±0.01[d]	0.05±0.01[c]	0.11±0.03[b]	0.14±0.02[b]	0.19±0.02[a]
Extractable Fe (ppm)	49.21±2.20[d]	67.19±5.87[c]	85.44±8.52[b]	90.64±9.15[ab]	98.67±1.50[a]
Extractable Cu (ppm)	0.14±0.03[e]	1.93±0.23[d]	4.93±1.10[c]	6.49±0.68[b]	9.42±0.95[a]
Extractable Zn (ppm)	0.94±0.65[d]	7.89±0.50[cd]	21.91±5.13[bc]	33.77±9.61[ab]	43.45±15.08[a]
Extractable Mn (ppm)	0.57±0.16[c]	4.98±1.04[c]	17.92±5.00[bc]	34.46±13.10[ab]	48.22±16.47[a]

*Means with different letter are significantly different at $p < 0.05$ according to LSD test.

The results showed a significant difference between the treatments (p<0.05), especially at the highest application rate of BT5 (Table 3). At harvest, the soil texture in BRIS soil changed to silt loam and sandy loam. Composting with chicken manure increased the percentage of clay and silt in the soils. Changes on the soil texture at harvest are resulted from increased composition of clay in the soil due to the ongoing bacteria cycle as well as decomposition processes (Ekhwan and Mazlin, 2009). Organic manure application was known to enhance soil porosity and water holding capacity and reduce soil compaction. It then increased the soil biopores and soil aeration, and better soil aggregation were observed (Mahmood et al., 2017). These factors lead to a better nutrient holding capacity and mineralization of nutrients to be uptake by the plant.

The soil pH was almost neutral in BRIS soil. The highest value of pH in soil after 9 months of treatment was observed in BT4 (6.88). Chicken manure has the ability to increase the pH of acidic soils (Basri et al., 2013). Most agronomic crops grow well at pH 6.0-7.0 (Khairi et al., 2011). The pH of the soil was observed to reduce after a few months of planting in all of the treatment. This is due to decarboxylation of organic anions due to decomposition and increased saturation for soil CEC by cations added by the manures (Carmo et al., 2015). During the decomposition process of the organic manure, organic acid and carbonic acid were produced and this resulted in lowering the soil pH (Chang et al., 1991).

Increasing trends of soil CEC with increased rate of chicken manure were observed with BT5 having the highest value of CEC (Table 3). Higher CEC were also observed in each treatment of soil at harvest compared to soil during the initial stage of planting. This concludes that, addition of chicken manure was proven to increase the CEC of treated soil and after a few months of its application, the CEC is improved more by the process of the decomposition and mineralization. However, BRIS soils, which have less soil colloids will have lesser CEC value due to leaching and base losses during rainy season (Basri et al., 2013). In addition, the presence of high clay and loam content in mineral soils helps to retain water for a certain time and helps prevent the leaching process of the nutrients from happening actively (Ekhwan and Mazlin, 2009).

Leaf nutrient characteristics
The application of chicken manure was observed to significantly affect the physicochemical properties in both soils. However, it has less effect on the nutrient content of the fig leaves (Table 4). Generally, N is essential for vegetative growth and reproductive development, P for fruit colour and maturation, and K for fruit yield and quality (Brunetto et a., 2015)). This is also applied to fig cultivation. In BRIS soil, leaf N and Ca did not show any significant difference at all treatments. Meanwhile, in mineral soil, leaf N and P did not show any significant difference at all treatments. Similar study by Han et al., (2016), soil was amended with compost from a mixture of poultry manure, cattle manure, swine manure, and sawdust, the N concentration was not changed in the leaves and stems by the treatment, but a significant increase was observed in the roots by the treatment.

Table 3: Physico-chemical properties of BRIS soil at harvest.

Treatment (v/v)	(0%)	(10%)	(20%)	(30%)	(50%)
Properties	**BT1**	**BT2**	**BT3**	**BT4**	**BT5**
Soil Texture	Silt Loam	Loam	Sandy Loam	Sandy Loam	Sandy Loam
pH	5.63 ± 0.04^e	6.45 ± 0.01^d	6.83 ± 0.01^b	6.88 ± 0.01^a	6.62 ± 0.01^c
C/N	4.00 ± 0.01^e	11.00 ± 0.01^a	9.56 ± 0.01^b	7.00 ± 0.01^d	9.23 ± 0.01^c
EC (dS/m)	0.11 ± 0.01^d	0.17 ± 0.01^{cd}	0.27 ± 0.04^c	0.42 ± 0.02^b	1.06 ± 0.16^a
CEC ($cmol_+$/kg)	1.44 ± 0.09^d	2.93 ± 1.51^{bc}	2.54 ± 0.13^{cd}	4.05 ± 1.18^b	7.05 ± 0.78^a
Total C (%)	0.12 ± 0.01^e	0.44 ± 0.01^d	0.86 ± 0.01^b	0.77 ± 0.01^c	2.03 ± 0.01^a
Total N (%)	0.03 ± 0.01^d	0.04 ± 0.01^{cd}	0.09 ± 0.02^{bc}	0.11 ± 0.01^b	0.22 ± 0.05^a
Extractable P %)	0.002 ± 0.003^d	0.016 ± 0.005^{cd}	0.032 ± 0.006^{bc}	0.046 ± 0.004^b	0.079 ± 0.017^a
Extractable K (%)	0.002 ± 0.001^b	0.005 ± 0.003^b	0.002 ± 0.001^b	0.004 ± 0.001^b	0.014 ± 0.006^a
Extractable Ca (%)	0.03 ± 0.02^d	0.19 ± 0.02^c	0.43 ± 0.13^b	0.51 ± 0.06^b	0.69 ± 0.09^a
Extractable Mg (%)	0.01 ± 0.01^b	0.01 ± 0.01^b	0.01 ± 0.01^b	0.04 ± 0.01^{ab}	0.13 ± 0.12^a
Extractable Fe (ppm)	41.28 ± 2.11^c	59.72 ± 7.64^{bc}	104.94 ± 30.42^a	78.45 ± 5.60^{ab}	89.17 ± 3.42^a
Extractable Cu (ppm)	0.11 ± 0.05^b	1.90 ± 0.19^b	5.87 ± 2.49^a	6.30 ± 2.31^a	9.04 ± 2.63^a
Extractable Zn (ppm)	3.32 ± 0.89^d	17.37 ± 0.46^c	35.19 ± 4.59^b	44.18 ± 4.04^b	64.03 ± 10.54^a
Extractable Mn (ppm)	0.43 ± 0.20^e	11.99 ± 1.85^d	22.24 ± 2.01^c	29.74 ± 1.63^b	43.26 ± 5.43^a

*Means with different letter are significantly different at $p < 0.05$ according to LSD test.

Table 4: Nutrients content of leaf samples at harvest.

Treatment (v/v)	(0%)	(10%)	(20%)	(30%)	(50%)
Properties	**BT1**	**BT2**	**BT3**	**BT4**	**BT5**
N (%)	3.42±0.53[a]	3.60±0.39[a]	3.18±0.09[a]	3.11±0.17[a]	3.15±0.07[a]
P (%)	0.14±0.02[b]	0.17±0.03[ab]	0.19±0.02[a]	0.17±0.03[ab]	0.15±0.01[ab]
K (%)	1.79±0.01[a]	1.48±0.01[bc]	1.66±0.27[abc]	1.75±0.18[ab]	1.41±0.14[c]
Ca (%)	1.19±0.07[a]	1.09±0.15[a]	1.22±0.09[a]	1.32±0.40[a]	1.20±0.10[a]
Mg (%)	0.16±0.02[b]	0.15±0.03[b]	0.18±0.03[ab]	0.21±0.03[ab]	0.24±0.04[a]
Fe (ppm)	394.93±227.58[a]	276.00±57.18[ab]	220.93±8.04[ab]	206.40±33.08[ab]	165.73±27.07[b]
Cu (ppm)	175.00±5.00[a]	142.50±2.50[ab]	164.17±41.26[a]	116.67±6.29[b]	115.00±0.01[b]
Zn (ppm)	552.50±265.32[a]	475.83±114.05[a]	440.83±29.19[a]	375.00±113.94[a]	443.33±105.37[a]
Mn (ppm)	176.67±46.00[c]	261.47±13.80[ab]	227.60±21.93[b]	286.00±22.03[a]	228.93±7.40[b]

*Means with different letter are significantly different at $p < 0.05$ according to LSD test.

Effectiveness of amendments of plant growth properties

The growth response of fig trees at harvest was shown in Table 5. The fig tree showed 100% of survivability in control and all of the soil treatments. The non-amended soils BT1, which is characterized by a slightly low pH were observed to have a significantly lowest growth. Acidic soils are unsuitable for fig production (Morton, 1987). Application of chicken manure was observed to significantly enhance the growth of fig trees through improvement of its physical and chemical properties. Fig grown on sandy soil requires regular and extensive fertilization. These practices had encouraged its growth for its fruit production (Stover and Aradhya, 2007). The highest plant height was recorded in BT3 (78.2 cm).

The treatment BT3 which received 20% of chicken manure showed a highest number of plant height (78.2 cm), stem girth width (3.22 cm), number of leaves (108 ± 31), number of branches (27 ± 8)·Fig trees are deciduous, fast growing, and spreading in habit, so that they tend to be greater in width than in height (Stover and Aradhya, 2007).

The number of fruits produced by BT4 (25) were recorded as the highest value. At the rate of 30%, fig trees were able to produce a significantly higher number of fruits compared to treatment that received 50% chicken manure (BT5). This could be due to excess fertilization that could promote excessive vegetative growth and results in low yield (Andersen and Crocker, 2013). This proved that chicken manure helps to improve the soil texture and water retention capacity hence increased the nutrient availability. Combination with organic manure in soil might have improved the N use efficiency, macro and micronutrients recovery and help in P solubilisation and its uptake by the plants and enhanced K availability that in turn resulted in better growth and yield (Mahmood et al., 2017). However, the rate has to be at optimum value so that the tree can perform well in terms of producing high yield.

BRIS soil amended with chicken manure is proven to achieve optimum growth and yield response. The most preferred soil for planting fig trees is the more balanced texture of amended BRIS soil, which helps in nutrient uptake for plant growth. Extensive area of BRIS soils in the coastal region of Peninsular

Malaysia can be economically important to be planted with figs along other crops. Amending BRIS soil with 30% of chicken manure is proven to be one of the best approaches to improve the soil health for fig planting as chicken manure is easily accessed and low in cost of agronomic practices.

Table 5: Average plant growth response of planted fig trees

Treatment (v/v)	(0%)	(10%)	(20%)	(30%)	(50%)
Properties	**BT1**	**BT2**	**BT3**	**BT4**	**BT5**
Survivability (%)	100	100	100	100	100
Plant Height (cm)	42.7 ± 13.4 [d]	75.3 ± 17.4 [ab]	78.2 ± 14.0 [a]	74.1 ± 17.5 [b]	69.3 ± 14.0 [c]
Stem Girth (cm)	1.57 ± 0.24 [d]	2.45 ± 0.75 [c]	3.22 ± 0.69 [a]	2.94 ± 1.07 [b]	2.33 ± 0.54 [c]
Number of Leaves	30 ± 12 [d]	77 ± 21 [c]	108 ± 31 [a]	97 ± 35 [b]	75 ± 26 [c]
Number of Branches	5 ± 2 [d]	19 ± 8 [c]	27 ± 8 [a]	23 ± 8 [b]	24 ± 11 [ab]
Number of Fruits	1 ± 1 [c]	16 ± 7 [b]	15 ± 5 [b]	25 ± 11 [a]	16 ± 12 [b]

*Means with different letter are significantly different at $p < 0.05$ according to LSD test.

CONCLUSION

Application of chicken manure on BRIS soil lead to significant changes in soil texture, pH, C/N ratio, EC, CEC, total C, total N, extractable P, K, Ca, Mg, Fe, Cu, Zn and Mn. The soil nutrient contents of control and treatments showed significant difference at ($p < 0.05$). However, the contents of nutrient in fig's leaf did not show any significant difference. For growth response and yield parameters, there were significant differences at ($p<0.05$) between the treatments. The survivability of the fig trees was 100% survived regardless of the rate of chicken manure used. Considering the economic aspect of fig planting, the best rate for chicken manure in BRIS soil would be 30% in order to achieve optimum growth and yield response. Amending BRIS soil with 30% of chicken manure can be one of the approaches to improve the soil health for fig planting as chicken manure is easily accessed and low in cost of agronomic practices. In addition, an extensive area of BRIS soils in the coastal region of Peninsular Malaysia can be economically important to be planted with figs along with other crops like tobacco, maize, kenaf and sweet potato.

ACKNOWLEDGEMENT

The authors would like to thank International Islamic University Malaysia (IIUM) for financial support, the Department of Plant Science and Department of Biotechnology at Kulliyyah of Science for technical support during the conduct of the research. This research was financed by the Research Initiative Grant Scheme (RIGS15-102-0102).

REFERENCES

Adenawoola, A. R. & Adejoro, S. A. (2005). Residual effects of poultry manure and NPK fertilizer residues on soil nutrients and performance of Jute (*Corchorus olitorius* L.). *Nigerian Journal of Soil Science.15*, 133-135.

Adeniyan, O. N. & Ojeniyi, S. O. (2005). Effect of poultry manure NPK 15:15:15 and combination of the reduced levels on maize growth and soil chemical properties. *Nigerian Journal of Soil Science. 15*, 34 – 41.

Andersen, P. C. & Crocker, T. E. (2013). The fig. *Institute of Food and Agricultural Sciences Extension.* 1-5.

Angelova, V. R., Akova, V. I., Artinova, N. S., & Ivanov, K. I. (2013). The effect of organic amendments on soil chemical properties. *Bulgarian Journal of Agricultural Science, 19* (5), 958-971.

Armanto, H. M. E., Adzemi, M. A., Wildayana, E. & Ishaq, U. M. (2013). *Coastal sand soils and their assessment for upland rice cultivation in Terengganu, Malaysia.* Proceedings of 2013 International Seminar on Climate Change and Food Security (ISCCFS 2013), Palembang, South Sumatra-Indonesia. Retrieved from http://eprints.unsri.ac.id/8027/ 1/Coastal_sand_Soils........pdf.

Awodun, M. A. (2007). Effect of poultry manure on the growth, yield and nutrient content of Fluted Pumpkin (*Telfaria occidentalis* Hook F.). *Asian Journal of Agricultural Research. 1*(2), 67-73.

Basri, M. H. A., Abdu, A., Jusop, S., Ahmed, O. H., Abdul-Hamid, H., Kusno, M. A., Zainal, B., Senin, A. L., & Junejo, N. (2013). Effects of mixed organic and inorganic fertilizers application on soil properties and the growth of kenaf (*Hibiscus cannabinus* L.) cultivated on BRIS soils. *American Journal of Applied Sciences. 10*(12), 1586-1597.

Bremner, J. M. & Mulvaney, C. S. (1982). Total N. In *Methods of Soil Analysis. 2nd edition* (pp. 600-601). Pages, A. L. (Ed.). Madison: American Society of Agronomy.

Carmo, D. L., Lima, L. B., & Silva, C. A. (2015). Soil fertility and electrical conductivity affected by organic waste rates and nutrient inputs. *Revista Brasileira de Ciencia do Solo.* 1-17.

Chang, C., Sommerfeldt, T. G. & Entz, T. (1991). Soil chemistry after eleven applications of cattle feedlot manure. *Journal of Environmental Quality. 20*, 475-480.

Davis, A. S., Jacobs, D. F., & Wightman, K. E. (2006). Organic matter amendment of fallow forest tree seedling nursery soils influences soil properties and biomass of sorghum cover crop. *Tree Planters' Notes.* 1-5.

Dikinya, O. & Mufwanzala, N. (2010). Chicken manure-enhanced soil fertility and productivity: Effects of application rates. *Journal of Soil Science and Environmental Management. 1*(3), 46-54.

Duncan, J. (2005). *Composting chicken manure.* WSU Cooperative Extension, King County Master Gardener and Cooperative Extension Livestock Advisor.

Ekhwan, M. T. & Mazlin, M. (2009). Analysis of the physical characteristics of BRIS soil in coastal Kuala Kemaman, Terengganu. *Research Journal of Earth Sciences. 1*(1), 1-6.

Ekhwan, M. T., Mazlin, M., Barzani, M. G., & Azlina, N. A. A. (2009). Analysis of the physical characteristics of BRIS soil in coastal Kuala Kemaman, Terengganu. *Research Journal of Earth Sciences. 1*(1), 1-6.

Eshoo, T. & Bell, L. C. (1992). Soil and solution phase changes and mugg bean response during amelioration of aluminium toxicity organic matter. *Plant Soil. 140*, 183-196.

Gustavo Brunetto, George Wellington Bastos De Melo, Moreno Toselli, Maurizio Quartieri, Massimo Tagliavini. (2015). The Role Of Mineral Nutrition On Yields And Fruit Quality In Grapevine, Pear And Apple. *Rev. Bras. Frutic., Jaboticabal.* SP, v. 37, n. 4, p. 1089-1104.

Han, S. H., An, J. Y., Hwang, J., Kim, S. B., & Park, B. B. (2016). The effects of organic manure and chemical fertilizer on the growth and nutrient concentrations of yellow poplar (*Liriodendron tulipifera* Lin.) in a nursery system. *Forest Science and Technology. 12*(2), 137-143.

Hue. N. V. & Amiens, I. (1989). Aluminium detoxification with green manures. *Communications in Soil Science and Plant Analysis. 20*, 1499-1511.

Ishaq, U. M., Armanto, H. M. E., & Adzemi, M. A. (2013). Performances of BRIS soils genesis and classification in Terengganu, Malaysia. *Journal of Biology, Agriculture and Healthcare. 3*(20), 86-92.

Ishaq, U. M., Umara, B., Armanto, H. M. E., & Adzemi, M. A. (2014). Assessment and evaluation of BRIS soil and its implication on maize crop in Merang-Terengganu region of Malaysia. *Journal of Biology, Agriculture and Healthcare. 4*(5), 69-76.

Kadir, W. R., Abdullah, R., & Ahmad, R. (2013). Application of organic residues for yield improvement of *Orthosiphon stamineus* on contrasting soils. Acta Biologica Indica. 2(1), 277-283.

Khairi, M. C. L., Nozulaidi, M. N., Musliania, M. I., Khanif, Y. M., & Sarwar, M. J. (2011). Composting increases BRIS soil health and sustains rice production. *Science Asia. 37*, 291-295.

Kidinda, L. K., Kasu-Bandi, B. T., Mukalay, J. B., Kabemba, M. K., Ntata, C. N., Ntale, T. M., Tamina, D. T., & Kimuni, L. N. (2015). Impact of chicken manure integration with mineral fertilizer on soil nutriments balance and Maize (*Zea mays*) yield: A case study on degraded soil of Lubumbashi (DR Congo). *American Journal of Plant Nutrition and Fertilization Technology.* *5*(3), 71-78.

Mahmood, F., Khan, I., Ashraf, U., Shahzad, T., Hussain, S., Shahid, M., Abid, M., & Ullah, S. (2017). Effects of organic and inorganic manures on maize and their residual impact on soil physico-chemical properties. *Journal of Soil Science and Plant Nutrition.* *17*(1), 22-32.

Malisa, M. N., Hamdan, J., & Husni, M. H. A. (2011). Yield response of kenaf (*Hibiscus cannabinus* L.) to different rates of charcoal and nitrogen fertilizer on BRIS soils in Malaysia. *Middle-East Journal of Scientific Research.* *10*(1), 54-59.

Mandal, M., Chandran, R. S., & Sencindiver, J. C. (2013). Amending subsoil with composted poultry litter-I: Effects on soil physical and chemical properties. *Agronomy.* *3*, 657-669.

Martin, R. C. & MacRae, R. (Eds.). (2014). *Managing energy, nutrients, an*

Mawa, S., Husain, K., & Jantan, I. (2013). *Ficus carica* L. (Moraceae): Phytochemistry, traditional uses and biological activities. *Evidence-Based Complementary and Alternative Medicine.* *2013*, 1-8.

McGovern, T. W. (2002). The fig – Ficus carica L. *Cutis.* *69*, 339-408.

Mehlich, A. (1984). Mehlich 3 soil test extractant: A modification of Mehlich 2 extractant. *Communication in Soil Science and Plant Analysis* 15, 1409-1416.

Nur Amirah, Y., Alias, A. A., & Wan Zaliha, W.S. (2015). Growth and water relations of roselle grown on BRIS soil under partial root zone drying. *Malaysian Applied Biology.* *44*(1), 63-67.

Ojeniyi, S. O., Amusan, O. A., & Adekiya, A. O. (2013). Effect of poultry manure on soil physical properties, nutrient uptake and yield of Cocoyam (*Xanthosoma saggitifolium*) in Southwest Nigeria. *American-Eurasian Journal of Agriculture and Environmental Science.* *13*(1), 121-125.

Okoli, P. S. O. & Nweke, I. A. (2015). Effect of different rates of poultry manure on growth and yield of Amaranthus (*Amaranthus cruentus*). *IOSR Journal of Agriculture and Veterinary Science.* *8*(2), 73-76.

Patil, V. V. & Patil, V. R. (2011). *Ficus carica* Linn.-An overview. *Research Journal of Medicinal Plant.* *5*(3), 246-253.

Pocknee, S. & Sumner, A. (1997). Cation and nitrogen contents of each of organic matter determines its liming potential. *Soil Science Society of America Journal.* *61*, 86-96.

Rhoades, J.D. (1982). Soluble salts. In: Page, A.L., R.H. Miller, and D.R. Keeney (Eds.). *Methods of soil analysis: Part 2: Chemical and microbiological properties. 2nd edition.* (pp. 167–179). ASA, Madison, WI.

Roslan, I., Shamshuddin, J., Fauziah, C. I., & Anuar, A. R. (2011). Fertility and suitability of the spodosols formed on sandy beach ridges interspersed with swales in the Kelantan-Terengganu plains of Malaysia for kenaf production. *Malaysian Journal of Soil Science.* *15*, 1-24.

Schollenberger, C. J. & Simon, R. H. (1945). Determination of exchangeable bases in soil- Ammonium acetate method. *Soil Science, 59*, 13-24.

Stover, E. & Aradhya, M. (2007). The fig: Overview of an ancient fruit. *HortScience.* *45*(5), 1083-1087.

Current Trend in Biotechnology and Molecular Biology of Fig (*Ficus carica* L.)

Muhamad Fahmi Yunus[1], Tamil Chelvan Meenakshi Sundram[1], Mohd Syafik Mohamad Hamdan[2], Nurul Hidayah Samsulrizal[1*]

[1]Department of Plant Science, Kulliyyah of Science, International Islamic University Malaysia (IIUM), 25200, Kuantan, Pahang, Malaysia.
[2] Bio Aromatic Research Centre of Excellence, Universiti Malaysia Pahang (UMP), 26300, Kuantan, Pahang, Malaysia.
**Corresponding author: hidayahsamsulrizal@iium.edu.my*

ABSTRACT

In vitro propagation offers some advantages over conventional propagation methods where it can supply uniform plants throughout the year and offer germplasm conservation. Besides, the optimization of *in vitro* culture conditions can accelerate and ease the research on transgenic plants through genetic engineering technology. Induced mutation breeding is a well-established method for plant improvement and this procedure can raise the possibility by a thousandfold when compared to spontaneous mutation under natural condition. Plant genetic engineering is currently a crucial method to transform genes of interest into a particular plant nuclear genome to get the desired expression. Successful transformation of fig cultivars provides a promising tool for the introduction of desired genes into fig cultivars, improved agronomic characteristic, and a means for the production of desired proteins in the edible parts of fig. Molecular markers are mostly neutral to environmental variations, in which researchers can evaluate their genetic material independently of the environmental conditions as opposed to morphological markers. Molecular markers such as RAPD, RFLP, ISSR and SSR have reportedly been prosperously utilized for the characterization of fig germplasm. Meanwhile, mutation which can occur either spontaneously or via induction plays a notable role in improving the ostiole size, fruit size, quality and productivity of fig. It is believed that fig has become one of the most valuable crops in the world as its fruits contain incredible medicinal properties and is used for the treatment of different ailments in traditional medicines.

Keywords: fig; genetic engineering; *in vitro* propagation; molecular markers; mutation

INTRODUCTION

Fig or scientifically known as *Ficus carica* L. is a deciduous plant belongs to Moraceae family. Prior to discovery of cereals, fig trees could have been the first domesticated plant in the Neolithic Revolution and has been reported to be domesticated five thousand years earlier compared to millet or wheat (Ciarmiello *et al.*, 2015). It is believed that fig is one of the first cultivated tree in the world due to its edible fruit. Figs are eaten raw as fresh fruits or processed into dried fruit or jam prior to consumption. Meanwhile in ethnobotany, the plant has been traditionally used as a natural laxative and treatment for gastrointestinal related illness such as colic, indigestion, constipation, and loss of appetite (Jagtap and Bapat, 2019).

Development of new cultivars is needed to fulfil the demand for high quality fruits. In 2014, the overall area covered by fig tree plantations around the world is approximately 426.244 hectares, with production reaching up to 1.070.676 million tonnes (Dessoky *et al.*, 2016). Most of the world's fig productions are concentrated in the Mediterranean basin which is well adapted to drought and high temperatures (Crisosto *et al.*, 2011; Barolo *et al.*, 2014).

Plant genetic improvement programme is one of the important fields in plant science which support the elements of food security. The genetic improvement technique provides tool to overcome problems regarding agricultural crops specially to meet world food supply. Nowadays, many agricultural systems have participated in the technology of genetically engineered crops, induced mutagenesis, and plant

55

tissue culture for mass production of plantlets. Those methods differ from conventional methods which require much less time compared to the traditional breeding techniques. This review seeks to discuss previous biotechnology research which include research on plant tissue culture, plant genetic engineering, mutation breeding, hairy root culture, molecular markers, molecular docking, and discussing the future of this valuable plant.

BIOTECHNOLOGY APPROACHES

Plant Tissue Culture

Fig seeds are non-viable and therefore the trees are often propagated by cutting, grafting and layering. However, this propagation method contributed towards synonymy and homonymy of the species , leading to misidentification of the plants (Kumar *et al.*, 1998). Moreover, this propagation approach is also less efficient, due to the phenotypic and genotypic variation of different mother plants that are being used (Ling *et al.*, 2018). Adding to this hindrance, Kumar *et al,* (1998), stated that due to the poor rooting, only 20-30% of the cuttings survive. Due to these problems, various alternative methods have been and are still being used to propagate fig plants.

Tissue culture (or known as *in vitro* propagation) is one of the biotechnological methods being used to propagate fig since the past two and a half decades. In general, through this method plants can be propagated at higher rates and controlled conditions in producing disease free plant Qrunfleh *et al.* (2013). Totipotency is the basic concept of plant tissue culture, with the understanding that any single cell can develop into differentiated cells and organ when provided with optimal conditions (Mat Taha, 2004).

Currently to our knowledge there are numerous studies of fig *in vitro* propagation using various different types of explants (i.e nodal segments, meristem and shoot tips) as stated in Table 1. Some of the main factors that contribute towards success of fig *in vitro* cultures are the cultivars, the explants used and the stages of the culture. Using technique known as somaclonal variation, three different cultivars with improved tolerance to salt have been produced i.e., Masone Black, Shami Stihy and Condaria cultivars (Aljane *et al.*, 2018). Some of the successful *in vitro* propagations of fig are summarized in Table 1.

Table 1: Summary of successful *in vitro* cultures of fig reported between 1998 and 2016.

Explants	Best medium treatment for shoot induction and multiplication of shoots	Best medium treatment for rooting and acclimatization	References
Shoot tips and nodal explants of Egyptian fig trees	Induction: Murashige & Skoog (MS) medium with 3.0 mg/L 6-benzylaminopurine (BAP) and 0.3 mg/L kinetin Multiplication: MS medium with 2.0 mg/L BAP, 0.1 mg/L kinetin and 0.1 mg/L gibberellic acid (GA$_3$) best for multiple shoots production	Rooting: Half-strength MS basal medium supplemented with 2.0 mg/L indole-3-acetic acid (IAA), 0.1 mg/L indole-3-butyric acid (IBA) and 2.0 g/L activated charcoal Acclimatization: 100% survival rate for acclimatized plants in pots containing a sterile soil under glasshouse condition	Dessoky *et al.* (2016)
Shoot tips of Conadria and Black Mission cultivars	Induction: MS medium supplemented with 0.5 mg/L BAP Multiplication: Full and double strengths MS medium increased the number of shoots and their length. Black Mission cultivar has highest number shoots while Conadria cultivar resulted in highest number of leaves and shoot length	Rooting: Half-strength MS medium with 0.1 M fructose was reported best for rooting	Taha *et al.* (2013)
Induction: Shoot tips were isolated from three-year-old Jordanian fig trees Multiplication: *In vitro* micro shoots, 15 mm in length	Induction: Half-strength MS medium Multiplication: MS medium with 0.4 mg/L BAP and 0.05 mg/L 1-naphthaleneacetic acid (NAA) Multiplication: Usage of sucrose, glucose or fructose as carbon sources resulted in variation in fresh and dry weights of *in vitro* shoots. However, no significant differences in the number of new shoots and shoot lengths. Increased NaCl (salinity) concentrations lowered the number of the newly formed shoots and shoot lengths	n/a	Qrunfleh *et al.* (2013)

Uniform shoot tip explants from different fig cultivars (Sultany, Aboudi and White Adcy)	Induction and multiplication: MS medium with 2.5 mg/L BAP resulted in highest number of shoot multiplication. MS medium supplemented with 2.5 mg/L 6-(γ,γ-dimethyl allylamino) purine (2iP) or adenine sulphate (ADS) resulted in highest number of leaves. Sultany cultivar responded the best	n/a	Mustafa and Taha, (2012)
Leaves from Seungjung Dauphine cultivar	Induction and multiplication: MS medium supplemented with either 0.5 mg/L or 1 mg/L thidiazuron (TDZ) with 2.0 mg/L IBA	Rooting: Full-strength MS basal medium with no PGR	Kim *et al.* (2007)
Sarilop cultivars clones 37, 50 and 82	Induction and multiplication: MS medium added with 1 mg/L IBA, 1 mg/L GA_3 and 5 mg/L BAP.	Rooting: MS medium with 0.75 mg/L and 1.5 mg/L IBA, or 0.67 mg/L and 1.3 mg/L NAA. Clone 37 is the best clone Acclimatization: Peat medium	Hepaksoy and Aksoy (2006)
Apical buds collected from 7- to 8-year-old cultivar Gular trees	Induction: MS medium with 2.0 mg/L BAP and 0.2 mg/L (NAA) Multiplication: MS medium with 2.0 mg/L BAP and 0.2 mg/L NAA	Rooting: Liquid half-strength MS medium supplemented with 2.0 mg/L IBA and 0.2% activated charcoal. Acclimatization: Survival rate in soil- 68% of the regenerated plantlets were successfully established in soil	Kumar *et al.* (1998)

n/a = not applicable

Somaclonal Variations, Mutation and Genetic Transformation

For decades, conventional breeding plays a vital role in crop development. Several approaches have been conducted using modern breeding techniques to develop fig with premium agronomic traits., The quality of crops can be improved through agriculture techniques by manipulating the genes of the desired properties (such as genes involved in resistance to insects, stress, and disease) (Table 2).

On the other hand, mutation breeding is another commonly used technique in breeding programme for crop improvement. Mutations are induced by using chemical mutagens like alkylating agents and colchicine, and also physical mutagens such as UV radiation, X rays, α- particles and β particles (Broertjes and Van Harten, 2013). In Turkey, the improvement of figs is mainly done by inducing mutations to fig trees in order to obtain bigger fruit, early ripening fruit and less ostiole size (Aljane *et al.*, 2018). According to Aljane *et al.* (2018) the new cultivar of Red Kadota has been produced through the mutation of Kadota.

Besides, genetic transformation is another method lately used in developing new cultivar. This process involves introduction and expression of foreign genes into a host organism. The overall efficacy of genetically modified plant techniques depends on (1) the efficiency of the transformation techniques used to stably insert the desired genes into plant cells or tissues, and (2) the regeneration technique(s) used to generate viable plants from transformed cells (Yancheva *et al.*, 2005). Moreover, it is a common approach used in molecular farming for the production of pharmaceutically important and commercially valuable proteins in plants. In molecular farming, the plants are genetically engineered to produce desired compounds such as vaccines, antibodies, therapeutic proteins and other industrial products such as protein and modified starches. The application of genetic engineering techniques used to integrate the homologous and/or heterologous genetic material into a species offers the potential of obtaining improved planting stocks in a short period of time (Yancheva *et al.*, 2005; Flaishman *et al.*, 2008).

Successful transformation of fig provides a promising tool for introducing desired genes into fig trees, improved agronomic characteristic, and means for the production of desired proteins in the edible parts of fig (Flaishman *et al.*, 2008). The efficient transformation protocols in fig have been developed with selective antibiotics, regeneration of transgenic shoots, molecular confirmation of transformants. Table 3 shows the fig plants that have already been manipulated using *Agrobacterium* mediated transformation technique.

Table 2: Somaclonal variations and mutations used in Fig traits improvement

Techniques	Cultivars tested	Results	References
Somaclonal variations	Masone Black, Shami Stihy, and Condaria	Salt tolerance increased. Number of leaves, shoots, fresh weight and dry weight as parameters	Aljane *et al.* (2018)
Spontaneous mutation through colchicine treatment	Moroccon Kadota	New cultivar developed i.e., Red Kadota with reduce ostiole size, increase fruit size, fruit quality and trees productivity	Flaishman *et al.* (2007)
Gamma radiation	Moroccon figs seed	New cultivar, 'Bol' with low tree vigour and rapid fruiting	Flaishman, *et al.* (2007)
Gamma radiation	Seed, cuttings and pollen of Egyptian figs	Dwarfness and rapid fruiting	Mars (2003)

Table 3: Efficient transformation and regeneration protocol of Fig

Cultivars tested	Description	Results	References
Black Mission	Molecular cloning and expression of a vacuolar Na+/H+ antiporter gene (*AgNHX1*) isolated from *Atriplex gmelina*	Putative regenerated transformant shoots containing *AgNHX1* gene were verified by PCR and Southern hybridization. Transgenic fig plants overexpressing *AgNHX1* gene grew normally in saline conditions compared to those of non-transgenic plants	Metwali *et al.* (2015)
Cultivar Sultani (fresh consumption)	*Agrobacterium* and microprojectile bombardment, the plasmid pISV2678 carrying the *gus*-intron and bar genes was used to transform the somatic embryogenesis explants	PCR analysis, histochemical GUS assay, and leaf painting assay demonstrated the regenerated plants' transgenic nature. Herbicide resistant plant established	Soliman *et al.* (2010)
Brown Turkey and Smyrna	1) *Agrobacterium* strain EHA105 harboring the plasmid pME504 that carried the *uidA*-intron, *bar* and *nptII* genes. 2) The Brown Turkey cultivar is transformed with grapevine (*Vitis vinifera* L.) cDNA encoding stilbene synthase and transcriptionally regulated by cauliflower mosaic virus (CaMV) 35S promoter to express the resveratrol compound which have several beneficial effects on health	1) The molecular confirmation through PCR and southern blot analysis using GUS reporter genes and BASTA resistance genes 2) Three transgenic plants of the cultivar were obtained, and resveratrol was identified in these transgenic plants. 3) Added nutritional value such as the resveratrol. Resveratrol provides health benefits to humans as it has anti-inflammatory, antiplatelet and anti-carcinogenic activities	Flaishman *et al.* (2008)
Cultivars of Brown Turkey (fresh consumption) and Smyrna (dry consumption)	*Agrobacterium* strain EHA105 containing plasmid pME504 that harbors the *uid*A-intron and *npt*II genes.	Molecular analyses and GUS staining validated the transgenic characteristic of the regenerated plants.	Yancheva *et al.* (2005)

Hairy Root Culture

Fig also produces various secondary metabolites, which do not only provide protection to plant but also useful for human's health. However, compounds obtain from wild- or field-grown plants are usually produced in trace amounts, which lead to the destruction of the natural habitat due to overexploitation of the plants (Rency *et al.*, 2019). Therefore, alternative biotechnological system by using hairy root culture is largely used for the mass production of secondary metabolites for medicinal plant resources (Rency *et al.*, 2019).

In hairy root culture technology, medicinal and herbal plants are genetically transformed to yield hairy roots which then will be scaled up in a bioreactor. Some of the applications of this technology includes production of secondary metabolites and novel compounds, germplasm conservation, phytoremediation, pharmaceuticals, molecular breeding and other industrial applications (Rency *et al.*, 2019). Hairy root cultures have gained its popularity in producing variety of valuable root-oriented metabolites including fig. Based on the study conducted by Amani *et al.* (2020), it is concluded that the hairy root culture system offers highly efficient *in vitro* technique for large scale production of phytochemical compounds from fig. Shoot and leaves explants from two fig cultivars (Siah and Sabz. two popular Iranian cultivars) were inoculated with different strains of *Agrobacterium rhizogenes* were utilized to establish the hairy root culture system for fig. Besides, elicitation of hairy roots of fig with methyl jasmonate (MeJA) was also conducted to obtain high value and yield of phenolic compounds and antioxidant contents in hairy roots. A maximum transformation rate of 100% was obtained by inoculating the shoots with *A. rhizogenes* strain A7. In addition, hairy roots of fig cultivar 'Siah' elicited with 100 and 200 µM MeJA and hairy roots of fig cultivar 'Sabz' elicited with 100 µM MeJA showed the highest total phenolic content. On top of that, MeJA elicitation substantially enhanced the content of six phenolic acids namely gallic acid, caffeic acid, chlorogenic acid, coumaric acid, cinnamic acid and rosmarinic acid, and three flavonoids which are rutin, quercetin, and apigenin (Amani *et al.*, 2020). As a result, induction of hairy roots and elicitation of fig hairy roots by MeJA showed a significant increase in production of phenolic compounds and antioxidants.

MOLECULAR BIOLOGY APPROACHES

Molecular Makers

Modern breeding programmes utilize the use of advanced molecular biology technology such as molecular markers. Different types of molecular markers are available for plant genetic diversity study, thus replacing the classically used conventional morphological markers. Major disadvantage of morphological markers is the plant phenotypic traits are easily influenced by environmental factors or phase of plant development. In contrast, the use of molecular markers are time saving and grant many advantages due to possibility of genomic DNA isolation from various plant tissues at any point of its growth. Targeted trait data can be acquired with linked DNA markers prior to pollination, enabling breeders to perform more informed genetic crosses. Furthermore, DNA markers are also regarded as a better option compared to traditional morphological markers due to the abundant, neutral, safe, easy to automate and cost-effectiveness (Kadirvel *et al.*, 2015). Some of the frequently used molecular markers are restriction fragment random amplified polymorphic DNA (RAPD), restriction fragment length polymorphisms (RFLPs), amplified fragment length polymorphisms (AFLPs), simple sequence repeats (SSRs) and single-nucleotide polymorphisms (SNPs). Four different types of molecular makers that have been used to detect the fig cultivars are summarized in Table 4.

Table 4: Molecular marker methods used for analysis of fig new cultivars

Molecular marker methods	Samples
ISSR	Jordanian Figs (Almajali *et al.,* 2012) and Chinese Fig (Zhang *et al.,* 2020)
SSR	Tunisian caprifigs (Essid *et al.,* 2015) and Spain fig (Perez-Jiménez *et al.,* 2012)
RAPD	Tunisian figs (Salhi-Hannachi *et al.,* 2006); Coruh Valley Figs, Turkey (Akbulut *et al.,* 2009); West Bank of Palestine Figs (Basheer-Salimia *et al.,* 2012), Dottato figs (Ciarmiello *et al.,* 2015) and Persia Figs, Iran (Baziar *et al.,* 2018)
RAPD and AFLP	Five new gamma irradiated plants were used in the study with Roxo-de-Valinhos cultivar as a standard (Rodrigues *et al.,* 2012)

According to Aljane *et al.,* (2018), ISSR and RAPD markers are mainly used to detect polymorphism and also in distinguishing fig cultivars. Meanwhile, SSRs procedures are habituated to identify, characterize, or for the assessment of genetic diversity of fig cultivars (Essid *et al.,* 2015; Aljane *et al.,* 2018). studies done by Rodrigues *et al.,* (2012) showed that RAPD and AFLP molecular markers have been used to compare the genetic variability gamma radiated fig trees and the Roxo-de-Valinhos cultivars. Other than that, Esmaiel *et al.* (2014) prosperously used EST analysis to twelve primer pairs of EST markers categorical for fig genus to investigate the genetic relatedness of 21 fig species.

Molecular Docking
Molecular docking approach refers to investigation on the interaction between molecules and proteins at the atomic level (Meng *et al.,* 2011). Initially fig compounds were gathered from the PubChem database and subjected to quantitative structure - activity relationship (QSAR) analysis and molecular docking, in order to reveal the compounds and proteasome interactions (Ayoub *et al.,* 2018). In the molecular docking approach carried out by Ayoub *et al.* (2018), the compound from fig was tested as proteasome inhibitors. Molecular docking is one of the most widely recognized structural approach, employed for molecular recognition studies in order to predict the binding mode and affinity of a complex formed by two or more constituent molecules with known structures (Roy *et al.,* 2015). In this case, the complex refers to ubiquitin–proteasome system (UPS) that is involved in various biological functions such as protein degradation and cell cycle regulation while the known structures refer to fig molecules being tested against the complex. Since the misregulation of proteasome was involved in cancer, targeting proteasome in an anticancer treatment is a promising approach for cancer therapy. Instead of utilizing drugs as inhibitors which showed potential side effects, fig molecules as proteasome inhibitors were investigated. The outcomes of QSAR modelling and docking have shown a great deal of potential for proteasomal inhibitors by fig compounds. For example, docking analysis demonstrated that catechin forms more non-covalent bonds in active site as compared to the reference drug Carfilzomib, due to its long molecules with large diameters, allowing for interaction and eventual inhibitor proteasome's active residues in its active site (Ayoub *et al.,* 2018)

CONCLUSION

Since fig is regarded as an important commercial plant, efforts in improving fig production cannot rely solely on conventional methods. Challenges such as land and water shortages, biotic and abiotic stress makes modern agricultural biotechnology and molecular biology techniques highly recommended for quality enhancement of fig. In molecular biology, hybridization, *Agrobacterium*-mediated transformation, *in vitro* mutagenesis are few examples of techniques that have been used for fig crop improvement. Moreover, various mutagenic agents have produced mutant fig with desired traits. There is also an urgent need to establish a germplasm bank that will provide information needed in future for genetic improvements of fig through active characterization and maintenance of fig varieties. For plant disease, fig mosaic disease is still a major problem in fig cultivation, which cause losses to the farmers. The usage of sequencing and nanotechnology can improve the fig crop for future needs. A genome editing method like CRISPR/Cas9 can be used on fig to generate non-GM fig fruits. In future, techniques such as protoplast fusion, biolistic transformation, microinjection, and microprojectile bombardment are methods that can be explored to mediate the genetic transformation procedure. These methods can be used to improve nutritional value of fig by introducing its precursor gene, namely stilbene synthase gene. As for postharvest of fig, improvising fruit storability and diseases resistance, enhancement of the nutritional and pharmaceutical constitution are research fields that can be explored. Morphological characters that can be improved include large fruit, early ripening fruit and small ostiole size that exhibit delayed ripening. Delayed ripening is needed as fig fruits are highly perishable and it makes postharvest handling hard.

FUTURE PERSPECTIVES

For future perspectives, plant molecular farming is a modern branch of plant biotechnology that can be explored for fig. Plants are modified to produce large amounts of recombinant pharmaceutical and industrial proteins with potential therapeutic or commercial values (Obembe *et al.*, 2011). High costs and inefficient production systems such as mammalian cells, transgenic animals and bacteria were reported to be unfavourable for the increasing quest as biomedicines which either be consumed as food or processed for topical application and health supplement (Moon *et al.*, 2020).

REFERENCES

Akbulut, M., Ercisli, S., & Karlidag, H. (2009). RAPD-based study of genetic variation and relationships among wild fig genotypes in Turkey. *Genetics and Molecular Research, 8*(3), 1109-1115. doi: 10.4238/vol8-3gmr634.

Alireza, T., & Nader, R. E. (2015). Molecular farming in plants. *Plants for Future; Intech Open: London, UK*, 25-41.

Aljane, F., Essid, A., & Nahdi, S. (2018). Improvement of fig (*Ficus carica* L.) by conventional breeding and biotechnology. In *Advances in Plant Breeding Strategies: Fruits* (pp. 343-375). Springer, Cham. doi: 10.1007/978-3-319-91944-7_9.

Almajali, D. A., Abdel-Ghani, A. H., & Migdadi, H. (2012). Evaluation of genetic diversity among Jordanian fig germplasm accessions by morphological traits and ISSR markers. *Scientia Horticulturae, 147*, 8-19. doi: 10.1016/j.scienta.2012.08.029.

Amani, S., Mohebodini, M., Khademvatan, S., & Jafari, M. (2020). *Agrobacterium rhizogenes* mediated transformation of *Ficus carica* L. for the efficient production of secondary metabolites. *Journal of the Science of Food and Agriculture, 100*(5), 2185-2197. doi: 10.1002/jsfa.10243.

Ayoub, L., Aissam, E.A., Yassine, K., Said, E., Mohammed, E.M., & Souad, A., (2018). A specific QSAR model for proteasome inhibitors from Oleaeuropaea and Ficuscarica. *Bioinformation, 14*(7), 384.

Barolo, M. I., Mostacero, N. R., & López, S. N. (2014). *Ficus carica* L. (Moraceae): an ancient source of food and health. *Food chemistry, 164*, 119-127. doi: 10.1016/j.foodchem.2014.04.112.

Basheer-Salimia, R., Awad, M., Salama, A., Alseekh, S., Harb, J., & Hamdan, Y. (2012). Molecular polymorphisms in Palestinian figs (*Ficus carica* L.) as revealed by random amplified polymorphic DNA (RAPD). *Journal of Genetic Engineering and Biotechnology, 10*(2), 169-175. doi: 10.1016/j.jgeb.2012.07.001.

Baziar, G., Jafari, M., Noori, M.S.S., & Samarfard, S. (2018). Evaluation of genetic diversity among Persian fig cultivars by morphological traits and RAPD markers. *HortScience, 53*(5), 613-619. doi: 10.21273/HORTSCI11306-16.

Broertjes, C., & Van Harten, A.M. (2013). *Applied mutation breeding for vegetatively propagated crops.* Elsevier.

Ciarmiello, L. F., Piccirillo, P., Carillo, P., De Luca, A., & Woodrow, P. (2015). Determination of the genetic relatedness of fig (*Ficus carica* L.) accessions using RAPD fingerprint and their agro-morphological characterization. *South African Journal of Botany, 97*, 40-47. doi: 10.1016/j.sajb.2014.11.012.

Crisosto, H., Ferguson, L., Bremer, V., Stover, E., & Colelli, G. (2011). Fig (*Ficus carica* L.). In *Postharvest biology and technology of tropical and subtropical fruits* (pp. 134-160e). Woodhead Publishing. doi: 10.1533/9780857092885.134.

Dessoky, E. S., Attia, A. O., & Mohamed, E. A. M. (2016). An efficient protocol for in vitro propagation of Fig (Ficus carica sp) and evaluation of genetic fidelity using RAPD and ISSR markers. *Journal of Applied Biology and Biotechnology, 4*(04), 57-63. doi: 10.7324/jabb.2016.40406.

Esmaiel, N. M., Abdellateif, K. F., Eldemery, S. M. M., Zakri, A. M., Al-Doss, A. A., & Barakat, M. N. (2014). Assessments of biodiversity of ornamental Ficus species based on EST markers and morphological traits. *Journal of Food, Agriculture & Environment, 12*(2), 932-938.

Essid, A., Aljane, F., Ferchichi, A., & Hormaza, J. I. (2015). Analysis of genetic diversity of Tunisian caprifig (*Ficus carica* L.) accessions using simple sequence repeat (SSR) markers. *Hereditas, 152*(1), 1-7. doi: 10.1186/s41065-015-0002-9.

Flaishman, M. A., Rodov, V., & Stover, E. (2008). The fig: botany, horticulture, and breeding. in Janick, J. (ed.) *Horticultural reviews.* 34th edn. New Jersey, USA: Wiley, pp. 113–197.

Flaishman, M. A., Yablovich, Z., Golobovich, S., Salamon, A., Cohen, Y., Perl, A., Yancheva, S.D., Kerem, Z & Haklay, E. (2008). Molecular breeding in fig (*Ficus carica*) by the use of genetic transformation. *Acta horticulturae, 798*, 151-158. doi: 10.17660/ActaHortic.2008.798.20.

Hepaksoy, S. E. R. R. A., & Aksoy, U. (2006). Propagation of *Ficus carica* L. clones by in vitro culture. *Biologia plantarum, 50*(3), 433-436. doi: 10.1007/s10535-006-0063-8.

Jagtap, U. B., & Bapat, V. A. (2020). Exploring Phytochemicals of *Ficus carica* L.(Fig). *Bioactive Compounds in Underutilized Fruits and Nuts*, 353-368. doi: 10.1007/978-3-030-06120-3_19-1.

Kadirvel, P., Senthilvel, S., Geethanjali, S., Sujatha, M., & Varaprasad, K. S. (2015). Genetic markers, trait mapping and marker-assisted selection in plant breeding. In *Plant biology and Biotechnology* (pp. 65-88). Springer, New Delhi. doi: 10.1007/978-81-322-2283-5.

Kim, K. M., Kim, M. Y., Yun, P. Y., Chandrasekhar, T., Lee, H. Y., & Song, P. S. (2007). Production of multiple shoots and plant regeneration from leaf segments of fig tree (*Ficus carica* L.). *Journal of Plant Biology, 50*(4), 440-446. doi: 10.1007/BF03030680.

Kumar, V., Radha, A., & Chitta, S. K. (1998). *In vitro* plant regeneration of fig (*Ficus carica* L. cv. Gular) using apical buds from mature trees. *Plant Cell Reports, 17*(9), 717-720. doi: 10.1007/s002990050471.

Ling, W.T., Liew, F.C., Lim, W.Y., Subramaniam, S., & Chew, B.L. (2018). Shoot induction from axillary shoot tip explants of fig (*Ficus carica*) cv. Japanese BTM6. *Tropical Life Sciences Research, 29*(2), 165-174. doi: 10.21315/tlsr2018.29.2.11.

Mat Taha, R. (2004). Kultur Tisu Tumbuhan Berbunga. Kuala Lumpur. Universiti Malaya Press.

Mars, M (2001), Fig (*Ficus carica* L.) genetic resources and breeding. In *II International Symposium on Fig 605* (pp. 19-27). doi: 10.17660/ActaHortic.2003.605.1.

Meng, X. Y., Zhang, H. X., Mezei, M., & Cui, M. (2011). Molecular docking: a powerful approach for structure-based drug discovery. *Current Computer-Aided Drug Design, 7*(2), 146-157.

Metwali, E. M., Soliman, H. I., Fuller, M. P., Al-Zahrani, H. S., & Howladar, S. M. (2015). Molecular cloning and expression of a vacuolar Na+/H+ antiporter gene (AgNHX1) in fig (Ficus carica L.) under salt stress. *Plant Cell, Tissue and Organ Culture (PCTOC), 123*(2), 377-387. doi: 10.1007/s11240-015-0842-z.

Moon, K. B., Park, J. S., Park, Y. I., Song, I. J., Lee, H. J., Cho, H. S., Jeon, J.H & Kim, H. S. (2020). Development of systems for the production of plant-derived biopharmaceuticals. *Plants, 9*(1), 30. doi: 10.3390/plants9010030.

Mustafa, N. S., & Taha, R. A. (2012). Influence of plant growth regulators and subculturing on *in vitro* multiplication of some fig (*Ficus carica*) cultivars. *Journal of Applied Sciences Research, 8*(8), 4038-4044.

Obembe, O. O., Popoola, J. O., Leelavathi, S., & Reddy, S. V. (2011). Advances in plant molecular farming. *Biotechnology Advances, 29*(2), 210-222. doi: 10.1016/j.biotechadv.2010.11.004.

Paul, M., & Ma, J. K. C. (2011). Plant-made pharmaceuticals: Leading products and production platforms. *Biotechnology and Applied Biochemistry, 58*(1), 58-67. doi: 10.1002/bab.6.

Perez-Jiménez, M., López, B., Dorado, G., Pujadas-Salvá, A., Guzmán, G., & Hernandez, P. (2012). Analysis of genetic diversity of southern Spain fig tree (*Ficus carica* L.) and reference materials as a tool for breeding and conservation. *Hereditas, 149*(3), 108-113. doi: 10.1111/j.1601-5223.2012.02154.x.

Qrunfleh, I. M., Shatnawi, M. M., & Al-Ajlouni, Z. I. (2013). Effect of different concentrations of carbon source, salinity and gelling agent on in vitro growth of fig (*Ficus carica* L.). *African Journal of Biotechnology, 12*(9). doi: 10.5897/AJB12.2871.

Rency, A. S., Pandian, S., Kasinathan, R., Satish, L., Swamy, M. K., & Ramesh, M. (2019). Hairy root cultures as an alternative source for the production of high-value secondary metabolites. In *Natural Bio-active Compounds* (pp. 237-264). Springer, Singapore.

Rodrigues, M. G. F., Martins, A. B. G., Desidério, J. A., Bertoni, B. W., & Alves, M. C. (2012). Genetic characterization of fig tree mutants with molecular markers. *Genetics and Molecular Research*, 1990-1996. doi: 10.4238/2012.August.6.3.

Roy, K., Kar, S., & Das, R. N. (2015). *Understanding the basics of QSAR for applications in pharmaceutical sciences and risk assessment*. Academic press.

Salhi-Hannachi, A. M. E. L., Chatti, K., Saddoud, O., Mars, M., Rhouma, A., Marrakchi, M., & Trifi, M. (2006). Genetic diversity of different Tunisian fig (*Ficus carica* L.) collections revealed by RAPD fingerprints. *Hereditas, 143*(2006), 15-22. doi: 10.1111/j.2005.0018-0661.01904.x.

Soliman, H. I., Gabr, M., & Abdallah, N. A. (2010). Efficient transformation and regeneration of fig (*Ficus carica* L.) via somatic embryogenesis. *GM crops, 1*(1), 40-51. doi: 10.4161/gmcr.1.1.10632.

Taha, R. A., Mustafa, N. S., & Hassan, S. A. (2013). Protocol for micropropagation of two *Ficus carica* cultivars. *world Journal of Agricultural sciences, 9*(5), 383-388. doi: 10.5829/idosi.wjas.2013.9.5.1802.

Yancheva, S. D., Golubowicz, S., Yablowicz, Z., Perl, A., & Flaishman, M. A. (2005). Efficient Agrobacterium-mediated transformation and recovery of transgenic fig (*Ficus carica* L.) plants. *Plant Science, 168*(6), 1433-1441. doi: 10.1016/j.plantsci.2004.12.007.

Zhang, X., Kong, W., Wang, X., Zhang, J., Liu, L., Wang, W., Zhang, H. & Deng, Q. (2020). Genetic diversity analysis of 34 fig varieties (*Ficus carica* L.) based on ISSR molecular marker. *Genetic Resources and Crop Evolution*, 1-9. doi: 10.1007/s10722-020-00889-5.

In Vitro Propagation of *Ficus* sp.

Zarina Zainuddin*
*Department of Plant Science, Kulliyyah of Science, International Islamic University Malaysia, Jalan
Sultan Ahmad Shah, Bandar Indera Mahkota, 25200 Kuantan, Pahang, Malaysia*
Corresponding author: zzarina@iium.edu.my

ABSTRACT

Ficus from the family Moraceae is a genus of around 850 species with many therapeutic and medicinal values. *Ficus* can be propagated via seeds, grafting, and cuttings. Realizing the importance of this genus and advantages of *in vitro* propagation can offers, many reliable methods for *in vitro* propagation have been developed resulting in rapid multiplication and mass production of clonal progenies. The present review emphasizes the *in vitro* manipulations of some important *Ficus* species mainly based on organogenesis, either directly or indirectly through callus formation. Regeneration of *Ficus* via somatic embryogenesis has also been reported for successful implementation of genetic transformation. Leaf is the most popular explant while shoot tip has also been used. Most of the works investigated on the influence of several factors including genotype, type, and concentrations of growth regulator, carbon source and type of media for successful micropropagation of *Ficus*.

Keywords: *in vitro propagation*, explant, organogenesis, somatic embryogenesis

INTRODUCTION

Ficus from the family of Moraceae is a diverse genus of more than 850 species (Harrison, 2005) consists of various forms: trees, shrubs, vines, epiphytes and hemiepiphytes (Somashekhar *et al.*, 2013). This genus is indigenous to the tropics (Choo and Sulong, 2016) and today it is cultivated in wide areas with the Asian-Australasian region having the richest and more diverse distribution (Badgujar *et al.*, 2014). *Ficus carica* is the most well-known *Ficus* species due to its commercial importance and applications (Lansky *et al.*, 2008). Other important species of *Ficus* include *Ficus deltoidea, Ficus religiosa, Ficus elastica, Ficus glomerata* and *Ficus benghalensis*. Ficus has many benefits to humans and is popular for its medicinal uses due to its phytochemicals that shows a high content of flavonoids, alkaloids, triterpenoids, coumarins, lactones and phenolics (Choo and Sulong, 2016). Humans consume dried figs since it has been claimed to contain multivitamins with minerals such as potassium, iron and calcium besides the high fibre content (Slavin, 2006). Due to the wide variety of chemical compositions in *Ficus*, this species has been shown to have biological activities of anti-cancer, anti-neoplastic, anti-oxidant and anti-inflammatory (Lansky *et al.*, 2008). Among the medical and therapeutical benefits of this species include the utilization of *F. carica* to treat illnesses such as constipation and eye vision problems (Abbasi *et al.*, 2013), colic and diarrhoea (Gilani *et al.*, 2008), kidney, urinary bladder stone, intestinal pain and dyspepsia (Marwat *et al*, 2009). Meanwhile for *F. deltoidea*, it has been used for treatment of diabetes, skin diseases, high blood pressure and to regulate the menstrual cycle (Choo *et al.*, 2012; Kiem *et al.*, 2013).

Propagation of *Ficus* plants is usually done by cuttings (Nava *et al.*, 2014). However, this propagation method can be time consuming. Seeds of fig can also be used to grow fig tree in a suitable condition (Ronsted *et al.*, 2008) but the germination rate is low (Rosnah *et al.*, 2015). Furthermore, *Ficus* species is difficult to distribute through direct seedling and usually distributes through animals (Harrison, 2005). Due to these problems and as the demand for *Ficus* is increasing, it is necessary to find alternative methods for its propagation, to ensure regular supply of its planting materials. Thus, instead of using traditional propagation, through *in vitro* propagation (or tissue culture), *Ficus* can be propagated in large quantities and provide continuous fresh materials to meet market demands for medicinal and pharmaceutical applications. Plant tissue culture is a propagation of plant parts or tissues that taken place in nutrient media (Bhoite and Palshikar, 2014). This type of propagation supports the production of shoots, roots and then followed by establishment after being transferred to soil. The development of tissue culture mainly depends on the concept of totipotency which refers to the ability of single cell to generate into entire plant with expression of full genome structure by cell division (Hussain *et al.*, 2012). Plant tissue culture is considered to be the most efficient technology for mass propagation, production of improved crop varieties, production of secondary metabolites, genetic transformation, production of disease-free plants and production of varieties tolerant to salinity, drought, and heat stress (Hussain *et al.*, 2012).

IN VITRO ESTABLISHMENT OF *Ficus* sp.

Among the *Ficus* species, *F. carica* is the most notably important. Thus, this species is the most studied in micropropagation through tissue culture. Besides *F. carica*, other species of *Ficus* where the tissue culture system has been developed includes *F. deltoidea, F. lyrata, F. glomerata, F. religiosa* and *F. benghalensis*. The establishment of *in vitro* culture of *Ficus* sp. has been achieved via both developmental pathways namely organogenesis and somatic embryogenesis. Formation of organs through organogenesis occurs either directly, without formation of callus or indirectly which callus is being formed first before organs formation is being induced. Similarly somatic embryogenesis in *Ficus* sp. can also be direct or indirect. Callus was induced from *Ficus* sp. to be used in establishing cell suspension culture for the production of useful secondary metabolites with commercial applications for example flavonoid compounds.

Callus initiation

Callus is actively dividing disorganized cell masses and is part of wound response (Ikeuchi *et al.*, 2013). Generally, in plant tissue culture, formation of callus can be induced using an intermediate ratio of auxin and cytokinin (Skoog and Miller, 1957). Callus has been initiated from *Ficus* sp. either using auxin and cytokinin alone or in combination of these two classes of plant growth regulators (PGRs). For example, callus started to form when leaves of *F. deltoidea* was cultured on Murashige and Skoog (MS) medium with as low concentration as 1.0 mg/l 6-benzylaminopurine (BAP) (Sa'adan and Zainuddin, 2020) while for *F. carica* 2.5 mg/l BAP was found to enhance callus formation (Mustafa and Taha, 2012). Besides BAP, another growth regulator from the class of cytokinin that is being used for callus initiation is Zeatin where 100% of semi-friable callus was formed when *F. carica* axillary shoot tips were cultured on medium with 0.5 – 2.0 mg/l Zeatin (Ling *et al.*, 2018). Among the growth regulators (picloram, indole-3-butyric acid; IBA and 2,4-dichlorophenoxyacetic acid; 2,4-D) tested for callus induction from leaves of *F. deltoidea*, 3.0 mg/l picloram was found to give the highest percentage of callus formation and to induce calli in the shortest time. The calli forms were friable and yellowish, greenish or white (Kiong *et al.*, 2007). Ismail *et al.* (2018) recorded the highest percentage of callus initiation when *F. deltoidea* leave explants were cultured on medium with combination of 1.5 mg/l picloram and 0.5 mg/l BAP. For the study of phytochemicals constituents in leaves and callus of *F. deltoidea* var. *Kunstleri*, Mustapha and Harun (2014) induced the callus on MS medium augmented with 0.8 mg/l 2,4-D + 0.7 mg/l kinetin. 2,4-D also has been used for callus initiation for *F. carica* where the best medium was found to be MS with 4.0 mg/l 2,4-D and 0.4 mg/l Kinetin (Danial *et al.*, 2014).

Shoot induction and multiplication

In general, addition of higher ratio of cytokinin to auxin in plant culture medium will result in formation of shoots (Hill and Schaller, 2013). BAP (or benzyl adenine; BA) is the most commonly used cytokinin for shoot induction, proliferation and multiplication in *in vitro* propagation of *Ficus* sp. either used singly (Sukamto, 1996; Danial *et al.*, 2014) or in combination with other cytokinin (Shahcheraghi and Shekafandeh, 2016) or auxin (Kumar *et al.*, 1998; Bayoudh *et al.*, 2015). The lowest BAP concentration that resulted in a high percentage of explants forming shoots was 0.5 mg/l, reported for regeneration of *F.*

glomerata (Hassan and Khatun, 2010). This is in accordance with results obtained by Rahman *et al.* (2004) on *in vitro* propagation of *F. benghalensis*. Other concentrations of BAP that gave the best or optimum results on shoot multiplication were 2.0 mg/l (Ling *et al.*, 2018), 2.5 mg/l (Mustafa and Taha, 2012) and 3.0 mg/l (Ripain *et al.*, 2015). BAP is routinely added in plant tissue culture media since it is highly stable and effective singly in over a wide range of plant species to induce shoot formation. For example, BAP alone resulted in the highest number of shoots for *in vitro* propagation of *Vernonia amygdalina* (Khalafalla *et al.*, 2007), *Ananas comosus* (Hamad and Taha, 2008), *Spilanthes acmella* (Haw and Keng, 2003) and *Hypericum retusum* (Namli *et al.*, 2010).

Hassan *et al.* (2009) and Mokhsin *et al.* (2008) combined BAP with indoe-3-acetic-acid (IAA) to achieve the best shoot induction for *F. religiosa* and *F. elastica*, respectively. Munshi *et al.* (2004) combined 1.0 mg/l BA with 0.1 mg/l 1-naphthaleneacetic acid (NAA) and 20% (v/v) coconut milk to produce large number of shoots in *F. benghalensis* while Deshpande *et al.* (1998) used the combination of 1.5 mg/l BA and 1.5 mg/l adenine sulphate (ADS). 5.0 mg/l BA has been combined with 1 mg/l gibberellic acid (GA$_3$) to enhance shoot multiplication in *F. carica* (Darwesh *et al.*, 2014) meanwhile Hepaksoy and Aksoy (2006) combined 5 mg dm^{-3} BA with 1 mg dm^{-3} gibberellic acid and 1 mg dm^{-3} IBA to achieve shoot multiplication.

Other cytokinins used for shoot induction in *in vitro* propagation of *Ficus* sp. include 2-isopentenyladenine (2ip), thidiazuron (TDZ) and kinetin. For example, shoot induction from callus of *F. carica* was achieved using MS medium augmented with 7.0 mg/l TDZ plus 0.5 mg/l NAA (Dhage *et al.*, 2012). In contrast, Kim *et al.* (2007) combined TDZ with IBA while Soliman *et al.* (2010) used a combination of TDZ with 2ip to achieve the best shoot multiplication and formation. TDZ is a synthetic phenylurea derivative (Khurana-Kaul *et al.*, 2010) which exhibits better shoot regeneration capacity in comparison to other cytokinins for a number of plant species (Radhika *et al.*, 2006; Pavingerová, 2009). Kinetin alone has been used for shoot proliferation in *F. carica* (Fráguas *et al.*, 2004; Pasqual and Ferreira, 2007) and *F. lyrata* (El-Agamy and Zeawail, 1991).

Root induction

In plant tissue culture, auxin plays the role in root organogenesis (Yu *et al.*, 2017). IBA is the most preferred auxin used for rooting of *Ficus* sp. incorporated either in full strength MS (Hepaksoy and Aksoy, 2006) or half strength MS (Bayoudh *et al.*, 2015; Shahcheraghi and Shekafandeh, 2016) medium. Rahman *et al.* (2004) obtained 100% of rooting when *in vitro* microcuttings of *F. benghalensis* were cultured on half strength of MS with 0.1 mg/l IBA. El-Agamy and Zeawail (1991) also obtained the same results for root formation in *F. lyrata* using MS medium supplemented with 0.5 mg/l IBA. Shoots of *F. benghalensis* rooted best on half MS medium fortified with 0.5 mg/l IBA (Munshi *et al.*, 2004). Some researchers achieved root induction by combining IBA with NAA. For example, Hassan and Khatun (2010) induced rooting in *F. glomerata* on half strength of MS + 2.0 mg/l IBA + 0.1 mg/l NAA while Deshpande *et al.* (1998) used full strength MS + 2.0 mg/l IBA + 0.1 mg/l NAA for *in vitro* propagation of *F. religiosa*. Compared to IAA, IBA is more stable in solution and in tissue, hence it has better ability to promote adventitious root formation (Ludwig-Müller, 2000).

Activated charcoal (AC) also has been combined with IBA for root induction in *Ficus* sp. Excised shoots of *F. carica* cv. gular were successfully rooted on half strength MS with 2.0 mg/l IBA and 0.2% AC (Kumar *et al.*, 1998). Similarly, the optimum medium for rooting for *F. anastasia* was found to be on half MS augmented with AC (3 g/l) and IBA or NAA (0.1 mg/l) (Al-Malki and Elmeer, 2009). Addition of AC to the rooting medium helps to adsorb inhibitory compounds and reduce the accumulation of toxic metabolites and phenolic exudates. The promoting effects of AC could be due to the darkening of the medium which mimics the soil conditions (Thomas, 2008).

Taha and co-workers (2013) investigated the effects of MS strength, type of carbon source and its concentration on rooting of two *F. carica* cultivars. It was found that half strength MS medium resulted in the highest percentage of rooting for both cultivars. Meanwhile, 0.1 M of fructose is the optimum carbon source and concentration for rooting compared to sucrose (0.1 M and 0.2 M). Best rooting response by half strength MS medium was also obtained by Al-Malki and Elmeer (2009) for root initiation of *F. anastasia*. The use of half MS resulted in highest root length and number of lateral roots compared to full strength MS. It was also observed in their studies that addition of 3 g/l of AC to half MS with either 0.1 mg/l IBA or NAA is the best for root initiation and induction. Half strength MS medium showed more effectiveness due to low concentration of nutrient compositions since relatively low salt concentrations in the growth medium is able to enhance adventitious rooting efficiency in shoots (Razdan, 2003). MS medium alone, without any addition of PGR has been reported to induce rooting in tissue culture of *F. carica* (Yakushiji *et al.*, 2003; Kim *et al.*, 2007).

Acclimatization

After *in vitro* plantlets/seedlings successfully rooted, they will be acclimatized prior to transferring to the field. This is to ensure a high survival rate when the plants are grown in the field. It was reported that acclimatization of *F. religiosa* was achieved by transferring rooted plants to plastic pots containing soil:compost (1:1), incubated in a polychamber with 80% relative humidity, temperature of 32 ± 2°C, and 12 hours light and 12 hours dark photoperiod (Hassan *et al.*, 2009). Munshi *et al.* (2004) also used a mixture of soil and compost for hardening of *F. beghaliensis* plantlets and resulted in about 80% of survival rate. Hassan and Khatun (2010) acclimatized plantlets of *F. glomerata* by keeping the plantlets in rooting cultures in normal room temperature prior to transplantation in pots. A study was conducted by Rahman *et al.* (2004) comparing the survival rate of *F. beghaliensis* plantlets when different potting mix was used for acclimatization. The highest plantlets survival rate (80%) was obtained when they were transferred to coco-peat while the lowest percentage (65%) was observed when soil and organic manure were used as potting mix.

Acclimatization of rooted plantlets of *F. carica* was tested using two different substrates, peat and peat+perlite. The best survival rate and plant growth were obtained on peat substrate with no abnormalities detected in the regenerated plants (Bayoudh *et al.*, 2015). Chirinéa *et al.* (2012) investigated the effect of time in culture medium (0, 15, 30, 45 and 60 days) and different substrates (Plantmax; Plantmax + Vermiculite [1:1 v/v]; Plantmax + Humus [1:1 v/v]; Humus + Vermiculite [1:1 v/v]; Plantmax + Humus + Vermiculite [1:1:1 v/v]) on plant height and survival rate during acclimatization. It was found that maintaining rooted plantlets in woody plant medium (WPM) for 30 days and acclimatized in Plantmax substrate are the optimum conditions for acclimatization in *F. carica*.

Somatic embryogenesis

Not many works have been conducted on somatic embryogenesis of *Ficus* sp. An efficient regeneration of *F. carica* via somatic embryogenesis (both indirect and indirect) was successfully developed by Soliman *et al.* (2010). Mitrofanova *et al.* (2019) also reported efficient morphogenesis in *F. carica* through somatic embryogenesis with successful percentage of 60-90% seedlings regeneration from somatic embryos. Somatic embryos formation was obtained on MS medium with 1.5 – 3.0 mg/l 2,4-D and 3.0 – 4.0 mg/l TDZ.

Suspension culture

Cell suspension culture system has been used mainly for production of valuable secondary metabolites and beneficial compounds from many plant species. For *Ficus* sp., most of suspension culture systems were developed for *F. deltoidea*. One example is a study conducted by Ong *et al.* (2011) on the effects of medium composition for the production of flavonoid compounds. Among the carbon sources tested, glucose was found to be the best sugar for production of rutin, quercetin and naringenin. Comparison on the effect of different auxins (2,4-D and NAA) on flavonoid production showed that the highest amount of rutin and naringenin was produced on medium with 10.74 µM NAA. Quercetin production was inhibited with addition of 2.4-D while 21.48 µM of NAA increased its production compared to control. It was also observed that supplementation of cytokinins (BAP and kinetin) suppressed production of rutin. This is in contrast to quercetin and naringenin where the highest amount was produced when 4.65 µM kinetin was added. Haida *et al.* (2019) investigated the effect of initial inoculum size, size of cell aggregate and initial pH on the cell biomass and production of antioxidant compounds in *F. deltoidea* var. *kunstleri* cell suspension culture. From the results it was observed that the highest production of biomass and flavonoid content were obtained from initial inoculum size of 2.0 g/25 ml media and 0.5 g/25 ml media, respectively. The cell aggregate size of 500 – 700 µm cultured at pH 5.75 are the optimum conditions for high cell biomass and flavonoid production. Another study on the optimization of growth parameters for cell culture of *F. deltoidea* was conducted by Ling *et al.* (2008). From their work, it was found that MS medium with 3 mg/l picloram is the best for establishment of cell suspension culture. Meanwhile fructose, inoculum size of 10 ml and 16 hours light and 8 hours dark photoperiod were the optimum conditions. Suspension culture was also successfully raised for *F. religiosa* for production of lectin. MS and WPM media supplemented with 10.95 µM 2,4-D or 4.4 µM BA + 10.95 µM 2,4-D were used to raise this suspension culture (Parasharami *et al.*, 2014).

FACTORS INFLUENCING *IN VITRO* PROPAGATION OF *Ficus* sp.

Selection of explants

Explants greatly influence the success of plant tissue culture. A successful culture can be established if explants are harvested from actively growing mother plant (George, 2008; Preece, 2008). Moreover, the response of explants to tissue culture conditions varies according to genetic and physiological determination (Fehér, 2006). Leaf segment is the most commonly used explant in *Ficus* micropropagation (Kim *et al.*, 2007; Mokhsin *et al.*, 2008, Dhage *et al.*, 2012). According to Yakushiji *et al.* (2003), for adventitious bud formation, leaf explant gave the results of 25% frequency of shoots/explant and 2.5 number of shoots/explants. Shoot tip is the second popular explant for *Ficus* shoot multiplication (Taha *et al.*, 2013; Ling *et al.*, 2018). Other explants being used for tissue culture of *Ficus* are node (Fráguas *et al.*, 2004), apical bud (Kumar *et al.*, 1998) and stem (Al Malki and Elmeer, 2010). Patah *et al.* (2018) tested four explants namely shoot, stem, root and petiole for callus induction of *F. carica*. The highest weight of callus was obtained from shoot (2.39 g), followed by stem explant (2.28 g) while most root and petiole were not responsive to hormone treatments. For regeneration of *F. glomerata*, it was found that nodal segments gave better response in terms of percentage of explant forming shoots and number of shoots/explants compared to shoot tips (Hassan and Khatun, 2010). Similarly in *F. religiosa*, nodal explants resulted in the highest frequency of callus induction compared to internodal segments and shoot apices (Siwach *et al.*, 2011).

Type of basal medium

Plant materials can grow successfully to new plantlets in each stage of *in vitro* clonal propagation when they get enough nutrients from medium and suitable physical conditions during the *in vitro* culture. The composition of the medium depends on the type of plant tissues or cells which vary from species to species. The most extensive used media for *in vitro* propagation of *Ficus* is MS medium either for formation of shoots (Deshpande *et al.*, 1998), roots (Kim *et al.*, 2007), callus (Ismail *et al.*, 2018) or establishment of suspension culture (Ling *et al.*, 2008). This basal medium contains all essential nutrient compounds needed for plant growth (Gamborg *et al.*, 1976; Bhojwani and Dantu, 2013) including inorganic nutrients (macronutrients, micronutrients), iron source and organic supplements which allow plants to grow (George *et al.*, 2008). It has been reported that MS medium allowed plant growth almost five to seven times more actively than other growth medium (Razdan, 2003).

Some researchers used different basal media for example Fráguas *et al.* (2004) used WPM to achieve shoot proliferation in *F. carica*. Pasqual and Ferreira (2007) also recommended the use of WPM for micropropagation of fig trees. Kumar *et al.* (1998) tested the response of apical bud of *F. carica* L. cv. gular towards shoot formation on three different media, MS, Gamborg B_5 and Shenk and Hildebrandt (SH). Although both MS and SH supplemented with 2.0 mg/l BAP + 0.2 mg/l NAA gave 100% of bud break, the number of shoot/explant is higher in MS (4.8) compared to SH medium (1.5). Meanwhile for B_5 medium, with the same PGRs and concentration, only 80% bud break was achieved and number of shoots/explant was 1.2. Study on the effect of using different type of medium was also done by Al-Shomali *et al.* (2007) where MS, WPM and Olive medium (OM) were used for callus development and shoot growth of three *F. carica* landraces (Khadri, Mwazi and Zraki). For Khadri, the use of three different media did not give significant difference on shoot growth but WPM was found to be the best for callus development. Meanwhile for Mwazi, the highest shoot growth and callus development were obtained from OM and WPM, respectively. No significant difference was observed in shoot growth of Zraki using the three media tested, while for callus development, the best medium was WPM.

Genotype

Genotypic dependency on *in vitro* propagation of *Ficus* sp. and differences in morphogenetic responses has been reported by a number of researchers. For example, Mustafa and Taha (2012) reported that among three cultivars of *F. carica* tested (Sultany, Aboudi and White Adcy), the highest number of callus formation, number of leaves and shoot length were obtained from Aboudi cultivar, meanwhile maximum shoot multiplication was obtained from Sultany cultivar. Dhage *et al.* (2012) tried to induce callus and regenerate plantlet from four different *F. carica* genotypes. It was shown that the Brown Turkey genotype produced the maximum number of callus and the only genotype that produced shoot compared to Conadria, Deanna and Poona. Another example of genotype dependency of *F. carica in vitro* propagation was shown by Hepaksoy and Aksoy (2006) where it was observed that clone 37 performed better than clone 50 and 82. Taha *et al.* (2013) compared the performance of two *F. carica* cultivars, where it was found that Black Mission gave better shoot number while Conadria resulted in higher leaves number and shoot length. Shoot initiation potential of three Tunisian local *F. carica* varieties (Zidi, Soltani and Bither Abiadh) was compared and results showed that Soltani has the highest potential (Bayoudh *et al.*, 2015).

Carbon sources

When explants are transferred to culture medium, they lack the photoautotrophic ability. Hence carbon source and energy in the form of sugar need to be included in the medium. Sucrose is the most common sugar used for *Ficus in vitro* propagation and added at the concentration of 3% (w/v) (Al Malki and Elmeer, 2009; Danial *et al.*, 2014; Ismail *et al.*, 2018). A higher concentration of sucrose, 4% was used for clonal propagation of *F. elastica* (Mokhsin *et al.*, 2008). Effect of adding lower concentration of sucrose ie. 2.5% and 2.0% for callus induction was conducted by Siwach *et al.* (2011). Addition of 2.5% sucrose did not differ significantly with 3.0% sucrose (control) while 2.0% resulted in lower callus induction. However, the fresh weight of callus obtained was found to be the same. In terms of the morphology of the callus, 3% and 2.5% sucrose produced compact and brownish green colour callus while watery, placid and pale-yellow colour callus was obtained from 2% sucrose. Mustafa *et al.* (2013) investigated the effect of carbon sources (sucrose and fructose) and their concentrations (0.1 and 0.2 mol/l) on shoot proliferation. In general, it was found that 0.2 mol/l fructose performed better than other sucrose treatments. Sucrose is the most commonly used in many *in vitro* propagation of plant species as it is cheap and easily found (Ahmad *et al.*, 2007). In fact, sucrose serves as a morphogenetic catalyst in the development of axillary buds and adventitious root branching (Vinterhalter and Vinterhalter, 1997).

71

Gelling agent and support matrix

The strength or concentration of gelling agents significantly influences the growth of cultured tissues. Agar is the most favourable gelling agent used to solidify *Ficus* culture medium and being added either at the concentration of 0.7% w/v (El-Agamy and Zeawail 1991; Mokhsin *et al.*, 2008; Bayoudh *et al.*, 2015) or 0.8% w/v (Siwach *et al.*, 2011; Darwesh *et al.*, 2014; Ling *et al.*, 2018). Rahman *et al.* (2004) used lower concentration of agar (0.6% w/v) for *in vitro* propagation of *F. benghalensis* while the lowest concentration of agar added ie 0.12% w/v was used for culture establishment of *F. carica* (Al-Shomali *et al.*, 2017). Agar (a polysaccharide from seaweeds) is the type of gelling agent which is most often used to prepare solid and semi-solid media as it assists in maintaining the matrix potential, humidity, water availability and the dissolved substances in containers (Debergh, 1983). Furthermore, agar produces no reaction when mixed together with other media components besides the amazing ability of not being digested easily by plant enzymes (Saad and Elshahed, 2012). Besides agar other gelling agents were also used. Gelrite was used to solidify culture medium for *in vitro* propagation of *F. deltoidea* (Kiong *et al.*, 2007; Ismail *et al.*, 2018) while phytagel is being added for tissue culture of *F. benghalensis* (Munshi *et al.*, 2004).

CONCLUSION

In vitro propagation or micropropagation is a platform for rapid regeneration of plantlets from explants on culture media supplemented with suitable plant growth regulators which overcomes the limitations of conventional methods that require longer time for propagation and multiplication. However, *in vitro* propagation depends on a number of factors such as environmental, physiological, nutritional and hormonal factors to achieve a successful regeneration of plantlets. Regeneration using stem cuttings of *Ficus* sp. takes longer time thus it is essential to establish *in vitro* culture which will rapid up the process and produce high yield of plantlets. In conclusion, the use of tissue culture for propagation of some economically important *Ficus* sp. acts as an alternative way to the conventional propagation techniques to meet the commercial demands. This technique has become one of the most prominent contribution of biotechnological advancements for the *Ficus* improvement programme. For decades, via micropropagation, a large number of desired *Ficus* plants have been regenerated under aseptic conditions in only a short period of time, with identical genetic composition, disease and pest-free as well as uniformly growing regardless of climatic and weather conditions.

REFERENCES

Abbasi, A. M., Khan, M. A., Khan, N., & Shah, M. H. (2013). Ethnobotanical survey of medicinally important wild edible fruits species used by tribal communities of Lesser Himalayas–Pakistan. *Journal of Ethnopharmacology, 148,* 528 – 536.

Ahmad, T., Abbasi, N. A., Hafiz, I. A., & Ali, A. (2007). Comparison of sucrose and sorbitol as main carbon energy sources in micropropagation of peach rootstock GF-677. Pakistan *Journal of Botany, 39*(4), 1269 – 1275.

Al Malki, A. A. H. S., & Elmeer, K. M. S. (2009). Effect of medium strength and charcoal combined with IBA and NAA on root initiation of *Ficus anastasia*. *Academic Journal of Plant Science, 2*(3), 169 – 172.

Al Malki, A. A. H. S., & Elmeer, K. M. S. (2010). Influence of auxin and cytokinine on *in vitro* multiplication of *Ficus anastasia*. *African Journal of Biotechnology, 9*(5), 635 – 639.

Al-Shomali, I., Sadder, M. T., & Ateyyeha, A. (2017). *Jordan Journal of Biological Sciences, 10*(1), 13 – 18.

Badgujar, S. B., Patel, V. V., Bandivdekar, A. H., & Mahajan, R. T. (2014). Traditional uses, phytochemistry and pharmacology of *Ficus carica*: A review. *Pharmaceutical Biology, 52*(11), 1487 – 1503.

Bayoudh, Ch., Labidi, R., Majdoub, A., & Mars, M. (2015). *In vitro* propagation of caprifig and female fig varieties (*Ficus carica* L.) from shoot tips. *Journal of Agricultural Science and Technology, 17,* 1597 – 1608.

Bhoite, H. A., & Palshikar, G. S. (2014). Plant tissue culture: A review. World Journal of *Pharmaceutical Sciences, 2*(6), 565 – 572.

Bhojwani, S. S., & Dantu, P. K. (2013). *Plant tissue culture: an introductory text.* New Delhi, Springer, India.

Chirinéa, C. F., Pasqual, M., De Araujo, A. G., Pereira, A. R., & De Castro, E. M. (2012). Acclimatization and leaf anatomy of micropropagated fig plantlets. *Revista Brasileira de Fruticultura, 34*(4), 1180 – 1188.

Choo, C-Y., & Sulong, N. Y. (2016). A review on the phytochemicals, ethnomedicine uses and pharmacology of Ficus species. *Current Traditional Medicine, 2,* 3 – 17.

Choo, C. Y., Sulong, N. Y., & Wong, T. W. (2012). Vitexin and isovitexin from the leaves of *Ficus deltoidea* with in vivo α-glucosidase inhibition. *Journal of Ethnopharmacology, 142,* 776 - 781.

Danial, G. H., Ibrahim, D. A., Brkat, S. A., & Khalil, B. M. (2014). Multiple shoots production from shoot tips of fig tree (*Ficus carica* L.) and callus induction from leaf segments. *International Journal of Pure and Applied Sciences and Technology, 20*(1), 117 – 124.

Darwesh, H. Y., Bazaid, S. A., & Abu Samra, N. (2014). In vitro propagation method of *Ficus carica* at Taif governate using tissue culture technique. *International Journal of Advanced Research, 2*(6), 756 – 761.

Debergh, P. C. (1983). Effects of agar brand and concentration on the tissue culture medium. *Physiologia Plantarum, 59*(2), 270 – 276.

Deshpande, S. R., Josekutty, P.C., & Prathapasenan, G. (1998). Plant regeneration from axillary buds of a mature tree *Ficus religiosa. Plant Cell Reports, 17,* 571 – 573.

Dhage, S. S., Pawar, B. D., Chimote, V. P., Jadhav, A. S., & Kale, A. A. (2012). *In vitro* callus induction and plantlet regeneration in fig (*Ficus carica* L.). *Journal of Cell and Tissue Research, 12*(3).

El-Agamy, S. Z., & Zeawail, M. El-F. (1991). *In vitro* propagation of *Ficus lyrata.* Meteorology, *Environment and Arid Land Agriculture Journal of King Abdulaziz University, 2,* 65 – 71.

Fehér, A. (2006). Why somatic plant cells start to form embryos? In Somatic Embryogenesis Plant Cell Monographs, eds A. Mujib and J. Šamaj, Heidelberg: Springer.

Fráguas, C. B., Pasqual, M., Dutra, L. F., & Cazetta, J. O. (2004). Micropropagation of fig (*Ficus carica* L.) 'Roxo De Valinhos' plants. *In Vitro Cell & Developmental Biology – Plant, 40,* 471 – 474.

Gamborg, O. L., Murashige, T., Thorpe, T. A., & Vasil, I. K. (1976). Plant tissue culture media. *In Vitro Cellular and Developmental Biology-Plant, 12*(7), 473 – 478.

George, E. F. (2008). Plant tissue culture procedure – Background. In: E. F. George, M. A. Hall & G. J. De Klerk (eds.), Plant propagation by tissue culture (3rd Edition) (pp. 403-422). Dordrecht, The Netherlands: Springer Science.

George, E. F., Hall, M. A. & De Klerk, G. J. (2008). The components of plant tissue culture media I: Macro- and micro-nutrients. In E. F. George, M. A. Hall & G. J. De Klerk (eds.), Plant propagation by tissue culture (3rd Edition), Dordrecht, The Netherlands: Springer.

Gilani, H. A., Mehmood, M. H., Janbaz, K. H., Khan, A., & Saeed, S. A. (2008). Ethnopharmacological studies on antispasmodic and antiplatelet activities of *Ficus carica. Journal of Ethnopharmacology, 119,* 1 – 5.

Haida, Z., Syahida, A., Mohd Ariff, S., Maziah, M., & Hakiman, M. (2019). Factors affecting cell biomass and flavonoid production of *Ficus deltoidea* var. kunstleri in cell suspension culture system. *Scientific Reports, 9,* 9533.

Hamad, A. M., & Taha, R. M., (2008). Effect of sequential subcultures on *in vitro* proliferation capacity and shoot formations pattern of pineapple (*Ananas comosus* L. Merr.) over different incubation periods. *Scientia Horticulturae, 117*(4), 329 - 334.

Harrison, R. D. (2005). Figs and the diversity of tropical rainforests. *BioScience, 55*(12), 1053 – 1064.

Hassan, A. K. M. S., Afroz, F., Jahan, M. A. A., & Khatun, R. (2009). *In vitro* regeneration through apical and axillary shoot proliferation of *Ficus religiosa* L. – A multi-purpose woody medicinal plant. *Plant Tissue Culture & Biotechnology, 19*(1), 71 – 78.

Hassan, A. K. M. S., & Khatun, R. (2010). Regeneration of *Ficus glomerata* Roxb., using shoot tips and nodal explants. *Bangladesh Journal of Botany, 39*(1), 47 - 50.

Haw, A. B., & Keng, C. L. (2003). Micropropagation of *Spilanthes acmella* L., a bio- insecticide plant, through proliferation of multiple shoots. *Journal of Applied Horticulture, 5*(2), 65 - 68.

Hepaksoy, S., & Aksoy, U. (2006). Propagation of *Ficus carica* L. clones by *in vitro* culture. *Biologia Plantarum, 50*(3), 433 – 436.

Hill, K., & Schaller, G. E. (2013). Enhancing plant regeneration in tissue culture. A molecular approach through manipulation of cytokinin sensitivity. *Plant Signaling & Behavior, 8*(10), e25709-1 – e25709-5.

Hussain, A., Qarshi, I. A., Nazir, H., & Ullah, I. (2012). Plant tissue culture: Current status and opportunities. In: Leva, A. & Rinaldi, L. (eds.), Recent Advances in Plant In Vitro Culture. IntechOpen.

Ikeuchi, M., Sugimoto, K., & Iwase, A. (2013). Plant callus: Mechanisms of induction and repression. *Plant Cell, 25*(9), 3159 – 3173.

Ismail, A., Shahidan, N., Othman, R., & Mat, N. (2018). Callus induction from leaf explants of mas cotek (Ficus deltoidea var. bilobata) treated with 6-benzylaminopurine (BAP) and 4-amino-3, 5, 6-trichloropicolinic acid (picloram). *International Journal of Engineering & Technology, 7*(4.43), 164 – 167.

Khalafalla, M. M., Elgaali, E. I., & Ahmed, M. M. (2007). *In vitro* multiple shoot regeneration from nodal explants of *Vernonia amygdalina* - an important medicinal plant. *African Crop Science Society Conference Proceedings, 8*, 747 - 752.

Khurana-Kaul, V., Kachhwaha, S., & Kothari, S. L. (2010). Direct shoot regeneration from leaf explants of *Jatropha curcas* in response to thidiazuron and high copper contents in the medium. *Biologia Plantarum, 54*(2), 369- 372.

Kim, K-M., Kim, M. Y., Yun, P. Y., Chandrasekhar, T., Lee, H-Y., & Song, P-S. (2007). Production of multiple shoots and plant regeneration from leaf segments of fig tree (*Ficus carica* L.). *Journal of Plant Biology, 50*(4), 440 – 446.

Kiem, P. V., Minh, C. V., Nhiem, N. X., Yen, P. H., Tuan Anh, H. I., Cuong, N. X., Tai, B. H., Quang, T. H., Hai, T. N., Kim, S. H., Kwon, S., Lee, Y., & Kim, Y. H. (2013). Chemical constituents of *Ficus drupacea* leaves and their a-glucosidase inhibitory activities. *Bulletin of the Korean Chemical Society, 34*, 263 - 267.

Kiong, A. L. P., Nee, T. C., & Hussein, S. (2007). Callus induction from leaf explants of *Ficus deltoidea* Jack. *International Journal of Agricultural Research, 2*(5), 468 – 475.

Kumar, V., Radha, A., & Kumar Chitta, S. (1998). In vitro plant regeneration of fig (*Ficus carica* L. cv. gular) using apical buds from mature trees. *Plant Cell Reports, 17*, 717 – 720.

Lansky, E. P., Paavilainen, H. M., Pawlus, A. D., & Newman, R. A. (2008). *Ficus* spp. (fig); Ethnobotany and potential as anticancer and anti-inflammatory agents. *Journal of Ethnopharmacology, 119*(2008), 195 - 213.

Ling, O. S., Kiong, A. L. P., & Hussein, S. (2008). Establishment and optimisation of growth parameters for cell suspension cultures of *Ficus deltoidea. American – Eurasian Journal of Sustainable Agriculture, 2*(1), 38 – 49.

Ling, W. T., Liew, F. C., Lim, W. Y., Subramaniam, S., & Chew, B. L. (2018). Shoot induction from axillary shoot tip explants of fig (*Ficus carica*) cv. Japanese BTM 6. *Tropical Life Sciences Research, 29*(2), 165 – 174.

Ludwig-Müller, J. (2000). Indole-3-butyric acid in plant growth and development. *Plant Growth Regulation, 32*, 219 – 230.

Marwat, S. K., Khan, M. A., Khan, M. A., Ahmad, M., Zafar, M., Fazal-ur-Rehman, & Sultana, S. (2009). Fruit plants species mentioned in the Holy Qura'n and Ahadith and their ethnomedicinal importance. *American-Eurasian Journal of Agricultural and Environmental Sciences*, *5*(2), 284 – 295.

Mitrofanova, I.V., Lesnikova-Sedoshenko, N.P., Brailko, V.A., Kuzmina, T.N., Chelombit, S.V., Shishkina, E.L., & Mitrofanova, O.V. (2019). Realization of *Ficus carica* L. morphogenic capacity via organogenesis and somatic embryogenesis *in vitro*. *Acta Horticulturae*, *1255*, 69 – 76.

Mokhsin, E. V., Lukatkin, A. S., & da Silva, J. A. T. (2008). Aseptic culture and simple, clonal micropropagation of *Ficus elastica* Roxb. *Floriculture and Ornamental Biotechnology*, *2*(2), 52 – 54.

Munshi, M. K., Hakim, L., Islam, M. R., & Ahmed, G. (2004). In vitro clonal propagation of banyan (Ficus benghalensis L.) through axillary bud culture. International Journal of *Agriculture & Biology*, *6*(2), 321 – 323.

Mustafa, N. S., & Taha, R. A. (2012). Influence of plant growth regulators and subculturing on *in vitro* multiplication of some fig (*Ficus carica*) cultivars. *Journal of Applied Sciences Research*, *8*(8), 4038 – 4044.

Mustafa, N. S., Taha, R. A., Hassan, S. A. M., & Zaied, N. S. M. (2013). Effect of medium strength and carbon source on *in vitro* shoot multiplication of two *Ficus carica* cultivars. *Journal of Applied Sciences Research*, *9*(4), 3068 – 3074.

Mustapha, Z., & Harun, H. (2014). Phytochemical constituents in leaves and callus of *Ficus deltoidea* Jack var. *Kunstleri* (King) Corner. *Walailak Journal of Science and Technology*, *12*(5), 431 – 439.

Namli, S., Akbas, F., Isikalan, C., Tilkat, E. A., & Basaran, D. (2010). The effect of different plant hormones (PGRs) on multiple shoots of *Hypericum retusum* Aucher. *Plant Omics Journal*, *3*(1), 12 - 17.

Nava, G. A., Júnior, A. W., Mezalira, E. J., Cassol, D. A., & Alegretti, A. L. (2014). Rooting of hardwood cuttings of Roxo de Valinhos fig (*Ficus carica* L.) with different propagation strategies. *Revista Ceres*, *61*(6), 989 – 996.

Ong, S. L., Ling, A. P. K., Poospooragi, R., & Moosa, S. (2011). Production of flavonoid compounds in cell cultures of *Ficus deltoidea* as influenced by medium composition. *International Journal of Medicinal and Aromatic Plants*, *1*(2), 62 – 74.

Parasharami, V., Yadav, P. Mandkulkar, S., & Gaikward, S. (2014). *Ficus religiosa* L.: Callus, suspension culture and lectin activity in fruits and *in vitro* regenerated tissues. *British Biotechnology Journal*, *4*(2), 215 – 227.

Pasqual, M., & Ferreira E.A. (2007) Micropropagation of fig tree (*Ficus carica* sp). In: Jain S.M., Häggman H. (eds) Protocols for micropropagation of woody trees and fruits. Springer, Dordrecht.

Patah, F. K. A., Hasbullah, N. A., Idris, H., Radzuan, N. S., & Mohd Lassim, M. (2018). Micropropagation of *Ficus carica* L. through tissue culture system. Proceedings of 12th International Conference on Advances in Agricultural, Chemical, Biological & Medical Sciences, Pattaya, Thailand.

Pavingerová, D. (2009). The influence of thidiazuron on shoot regeneration from leaf explants of fifteen cultivars of *Rhododendron*. *Biologia Plantarum*, *53*, 797 – 799.

Preece, J. (2008). Stock plant physiological factors affecting growth and morphogenesis. In: E. F. George, M. A. Hall & G. J. De Klerk (eds.), Plant propagation by tissue culture (3rd Edition) (pp. 403-422). Dordrecht, The Netherlands: Springer Science.

Radhika, K., Sujatha, M., & Rao, T. N. (2006). Thidiazuron stimulates adventitious shoot regeneration in different safflower explants. *Biologia Plantarum*, *50*, 174 - 179.

Rahman, M.M., Amin, M. N., & Hossain, M. F. (2004). *In vitro* propagation of banyan tree (*Ficus benghalensis* L.) – A multipurpose and keystone species of Bangladesh. *Plant Tissue Culture*, *14*(2), 135 – 142.

Razdan, M. K. (2003). Introduction to plant tissue culture. Enfield, USA: Science Publishers, Inc.

Ripain, I. H. A., Vikram, P., Kayat, F., Susanto, D., & Aurifullah, M. (2015). In vitro shoot regeneration of Mas Cotek (*Ficus deltoidea* Jack) a valuable Malay medicinal plant. Journal of *Tropical Resources and Sustainable Science*, *3*, 46 – 50.

Ronsted, N., Weiblen, G.D., Clement, W.L., Zerega, N.J.C., & Savolainen, V. (2008). Reconstructing the phylogeny of figs (Ficus, Moraceae) to reveal the history of the fig pollination mutualism. *Symbiosis, 45*, 45 - 55.

Rosnah, J., Khandaker, M. M., & Boyce, A. N. (2015). *Ficus deltoidea*: A review on background and recent pharmacological potential. *Journal of Agronomy, 14*(4), 310 – 318.

Sa'adan, H., & Zainuddin, Z. (2020). Callus induction from leaf explant of *Ficus deltoidea* var Kunstleri. *Science Heritage Journal, 4*(1), 06-08.

Saad, A. I., & Elshahed, A. M. (2012). Plant tissue culture media. In: Recent Advances in Plant In Vitro Culture. IntechOpen.

Shahcheraghi, S. T. & Shekafandeh, A. (2016). Micropropagation of three endemic and endangered fig (*Ficus carica* L.) genotypes. *Advances in Horticultural Science, 30*(3), 129 – 134.

Siwach, P., Gill, A. R., & Kumari, K. (2011). Effect of season, explants, growth regulators and sugar level on induction and long term maintenance of callus cultures of *Ficus religiosa* L.. *African Journal of Biotechnology, 10*(24), 4879 – 4886.

Skoog F., & Miller C.O. (1957). Chemical regulation of growth and organ formation in plant tissues cultured *in vitro*. *Symposia of the Society for Experimental Biology, 11*, 118–130.

Slavin, J. L. (2006). Figs: Past, present, and future. *Nutrition Today, 41*(4). 180 – 184.

Soliman, H. I., Gabr, M., & Abdallah, N. (2010). Efficient transformation and regeneration of fig (*Ficus carica* L.) via somatic embryogenesis. *GM Crops, 1*(1), 40 – 51.

Somashekhar, M., Nayeem, N., & Mahesh, A. R. (2013). Botanical study of four Ficus species of family Moraceae: A review. *International Journal of Universal Pharmacy and Bio Sciences, 2*(6), 558 – 569.

Sukamto, L. A. (1996). Micropropagation of *Ficus benjamina* L. 'Variegata'. *HortScience, 31*(3), 324b – 324.

Taha, R. A., Mustafa, N. S., & Hassan, S. A. (2013). Protocol for micropropagation of two *Ficus carica* cultivars. *World Journal of Agricultural Sciences, 9*(5), 383 – 388.

Thomas, T. D. (2008). The role of activated charcoal in plant tissue culture. *Biotechnology Advances, 26*(6), 618 – 631.

Vinterhalter, D. & Vinterhalter, B. (1997). Micropropagation of Dracaena species. In High-Tech and Micropropagation VI (pp. 131-146). Springer, Berlin, Heidelberg.

Yakushiji, H., Mase, N., & Sato, Y. (2003) Adventitious bud formation and plantlet regeneration from leaves of fig (*Ficus carica* L.). *The Journal of Horticultural Science and Biotechnology, 78*(6), 874 - 878.

Yu, J., Liu, W., Liu, J., Qin, P., & Xu, L. (2017). Auxin control of root organogenesis from callus in tissue culture. *Frontiers in Plant Science, 8*, 1385.

Genome Size Determination of *Ficus deltoidea* Using Flow Cytometry Method

Siti Sarah Zailani[1], Muhammad Irfan Fikri[1] & Mohd Razik Midin[1*]
[1]Department of Plant Science, Kulliyyah of Science, International Islamic University Malaysia, Jalan Sultan Ahmad Shah, Bandar Indera Mahkota, 25200 Kuantan, Pahang, Malaysia
Corresponding author: mohdrazik@iium.edu.my

ABSTRACT

Ficus deltoidea is a large shrub native to Southeast Asia. Locally, it is known as 'Mas Cotek'. It is famous for its medicinal value and health benefits. As for now, genome size data of *F. deltoidea* is not available yet. This study aims to optimise nuclei preparation method of flow cytometry (FCM) measurement as well as determine the nuclear genome size of *F. deltoidea*. Several steps involved in this study: (1) isolation of nuclei using two different techniques (manual chopping and Medi-machine) and two different lysis buffers (Otto and LBO1), (2) nuclear DNA staining using propidium iodide (PI) dye and (3) measurement of fluorescent intensity of stained nuclei. DNA histograms generated from different methods and lysis buffers were compared. This study revealed that the best DNA histogram was obtained by using manual chopping technique supplemented with Otto buffer. Using FCM with *Glycine max* cv. Polanka (2C=2.5pg) as an external reference standard, the genome size of *F. deltoidea* was found to be 2.88 pg. The genome size data can be utilised in future studies of ecology, evolution, taxonomy as well as genome sequencing.

Keywords: *Ficus deltoidea,* flow cytometry, genome size

INTRODUCTION

Ficus deltoidea is a perennial herb that native to Malaysia. It is widely found throughout Thailand, Sumatra, Java, Kalimantan, Sulawesi and Moluccas. Locally, it is known as mistletoe fig or "Mas cotek" due to the golden-yellowish colour on the upper surface of the leaf (Hasan et al., 2012). The leaf morphology of this species is greatly variable (Kochummen and Rusea, 2000).

F. deltoidea has economic value due to its medicinal properties (Zakaria et al., 2012). It contains a great source of antioxidant and natural products (Rosnah et al., 2015). Its leaves and fruits can be exploited for medicinal purposes such as post-natal, hypertension and diabetes treatments (Hakiman and Maziah, 2009; Adam et al.,2011; Adam et al., 2012; Rosnah et al.,2015). Previous studies found that the leaves of *F. deltoidea* exhibit blood glucose-lowering effects and possess antioxidant activity (Abdulla et al., 2010; Rosnah et al., 2015). Traditionally, its fruits are chewed to relieve toothache and headache (Rosnah et al., 2015). In addition, its leaves, stems and roots are dried and commercialised as herbal tea in Malaysia (Subhash et al., 2009). Due to its potential values, *F. deltoidea* is grouped under Entry Point Project 1 (EPP1) of National Key Economic Area (NKEA) Agriculture (Ripain et al., 2015).

Most studies conducted on *F. deltoidea* are related to the phytochemistry and taxonomy (Sulaiman et al., 2008; Abdullah et al., 2009; Adam et al., 2012; Bunawan et al., 2014). Up to now, the genome size of *F. deltoidea* is still not available. Genome size is defined as the total amount of DNA in the nucleus. It can be measured either in picograms or megabase pairs (Pellicer et al., 2018). Genome size data can be used to predict the cost and time for genome sequencing, species delimitation and determine the ploidy manipulation in plant breeding as well as tissue culture (Albach and Greilhuber 2004; Bennett and Leitch, 2005; Vrána et al. 2014).

Currently, high-throughput flow cytometry (FCM) is widely used in genome size determination. It has become the method of choice as it is a fast and efficient method which that requires small tissues from samples (Doležel and Bartoš, 2005). In the previous studies, FCM method has been applied in plant breeding, plant pathology, plant reproductive biology and other advanced plant science research (Brown,

1994; Ochatt, 2008; D'Hondt et al., 2011; Hojsgaard and Herbig, 2013). Prior FCM measurement, the nuclei suspension preparation should be optimised by using different lysis buffers and nuclei isolation techniques. Due to this, the aims of this study are to optimize the nuclear suspension method as well as determine the genome size of *F. deltoidea.*

MATERIALS AND METHODS

Plant material
The plant samples were collected from Glasshouse Nursery Complex (GNC) in International Islamic University Malaysia, Kuantan Campus, Pahang. Young leaf tissues of *F. deltoidea* (Figure 1A) were utilized for FCM analysis as they are easier to handle and contain fewer secondary metabolites (Jedrzejczyk and Sliwinska, 2010; Midin et al., 2017). *Glycine max* cv. Polanka (soybean) (Figure 1B) was used as an external reference standard in the FCM analysis.

Fig. 1: Young plant of *F. deltoidea* (A) and soybean (B)

Nuclei extraction and staining
The nuclei suspension was extracted from young leaf samples using two different techniques (medimachine blender and manual chopping with razor blade) and nuclei isolation buffers namely LBO1 and Otto buffer. The extracted nuclei suspensions were then filtered using 50 μm nylon mesh into a 10 mL falcon tube (Becton Dickinson, USA). The released nuclei suspensions were supplemented with RNase A (Sigma-Aldrich, USA) and Propidium Iodide (PI) (Sigma-Aldrich, USA) prior analysis.

Genome size determination
Stained nuclei were analysed using FCM-FACSCalibur machine (BD Biosciences, San Jose, CA) supplied with 15 mW argon ion laser at 488 mm. CellQuest software integrated with FCM machine generated DNA histogram peak based on the fluorescence intensity emitted by each stained nucleus. The machine was setup to capture 3000 nuclei for each sample. For genome size calculation, *Glycine max* cv. Polanka (Soybean, 2C=2.5 pg) was used as an external reference standard. Fluorescence intensity of targeted samples were compared with fluorescence intensity of soybean in order to obtain the ratio, then, multiplied by the genome size of soybean, 2C=2.5 pg.

RESULTS AND DISCUSSION

Optimising nuclei suspension preparation

DNA histogram peak generated by FCM is an important indicator to select the suitable nuclei isolation buffer prior to genome size estimation. Good histogram DNA peak must fulfil all these three conditions: (1) minimal debris background, (2) symmetrical DNA peak, and (3) low CV value (Dolezel & Bartos, 2005).

This study found that the suitable nuclei isolation buffer for *F. deltoidea* was Otto. Based on the result, Otto (Figure 2A & 2B) generated a good histogram DNA peak as compared to LBO1 (Figure 2C & 2D). The DNA peak produced by Otto demonstrated less debris and narrow DNA peak with low CV value (<3%) at channel 160-280. This result proved that different lysis buffers give different effects on the stability of nuclei and DNA staining (Vrána et al., 2014). Loureiro et al. (2006) speculated that LB01 and Otto are excellent lysis buffers. However, Otto buffer offers better results in many plant species especially those with lower nuclear genome size.

The chemical components of buffers can affect the result of FCM. In LBO1, the composition of EDTA acts as a metal chelator that will bind to nucleases cofactor (or divalent cation). The presence of inorganic salts such as NaCl and KCl provides sufficient ionic strength in the buffer. The pH for both Otto and LB01 buffers is adjusted around 7.5, well-suited for DNA staining. In addition, the presence of non-ionic detergents such as Triton X-100 and Tween 20 facilitate the release of nuclei and reduce the aggregation affinity of nuclei and debris. Other compounds such as $MgCl_2$, $MgSO_4$, and spermine will stabilize the chromatin (Doležel et al., 2007). As for now, there is no specific buffer suitable for all plant species (Loureiro et al., 2006). Due to this, the selection of a suitable lysis buffer during nuclei suspension preparation is a necessary step prior genome size determination.

Other factors that will affect the quality of nuclei suspension is the nuclei isolation technique. This study found that the combination of manual chopping and Otto buffer (Figure 2A) produced fine DNA histogram peak with less debris background DNA. The debris is mainly found on the right and lower fluorescent side of all histograms. High content of debris background can be caused by excessive chopping and selection of unsuitable buffers. Manual chopping avoids the damage of nuclei that can reduce the amount of debris (Doležel et al., 2007). Medimachine is an automated tissue blender that allows the uncontrolled speed during tissue blending. Consequently, the nuclei will damage and yield debris peaks on histogram. Over chopping using scalpel will give the same errors (Doležel et al., 2007). The use of blunt razor blade or scalpel contributes to the presence of background noise in histogram (Doležel et al., 2007). Among all existing nuclei isolation techniques (bead beating, automated tissues blending) manual chopping has been reported as the best method to isolate nuclei from leaf tissues (Roberts, 2007; Doležel and Greilhuber, 2010) and this has been proven in this study. Out of all histograms, Figure 2A presents the best histogram DNA peak indicating that manual chopping technique using Otto lysis buffer can be used for genome estimation of *F. deltoidea*. Table 1 summarises CV value from all four preparation methods used.

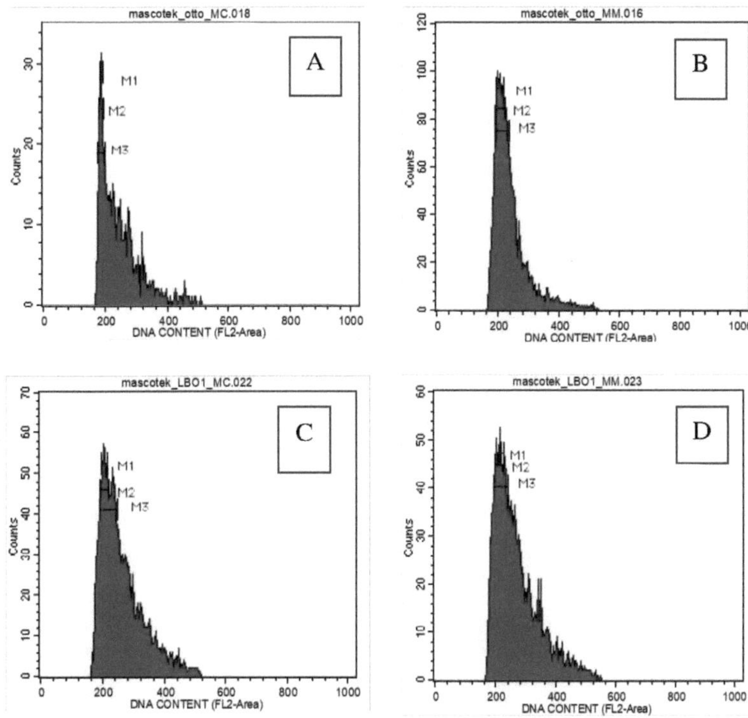

Fig. 2: Histogram DNA peak of *F. deltoidea*:(A) Otto-Manual chopping, (B) Otto-Medimachine, (C) LBO1-Manual chopping and (D) LBO1-Medimachine

Table 1: CV value obtained by all four nuclei preparation methods

Histogram	Preparation methods	Average CV
A	Otto-Manual chopping	2.31
B	Otto-Medimachine	3.87
C	LBO1-Manual chopping	4.85
D	LBO1-Medimachine	4.38

Genome size determination

Genome size determination of *F. deltoidea* was carried out on the fine histogram DNA peak (with CV less than 3%) using *Glycine max* cv. Polanka (soybean, 2C = 2.5 pg) as an external reference standard. Soybean was utilised as an external reference standard because; (1) it has established genome size (2.5 pg) and (2) manageable during sample preparation (Madon et al., 2008). External standardisation was applied in this study following the Hendrix and Steward (2005) method to yield non-overlapping DNA peaks.

Three conditions must be achieved to generate relevant genome size data including (1) sufficient amounts of isolated nuclei with stable DNA molecule, (2) the fluorochrome used must be specific and stoichiometric to both DNA sample and standard and (3) the use of suitable and known genome size of reference standard (Doležel and Bartoš, 2005). In this study, all these conditions were achieved. Thus, the genome size of *F. deltoidea* was successfully estimated. This is the first report on the genome size of *F. deltoidea*. The genome size was found to be 2.88 pg (2,817 Mbp, 1 pg=978 Mbp) (Table 2).

Table 2: Genome size of *F. deltoidea* (prepared using manual chopping and Otto buffer) using soybean (2C=2.5 pg) as an external reference standard

G1 peak mean of each replicate (5 replicates)	G1 peak mean of soybean	Genome size of *F. deltoidea* (pg)
223.41	193.77	2.88

CONCLUSION

In summary, the genome size of *F. deltoidea* was successfully measured by the FCM method. The data can be utilized as a reference for future research. Intraspecific genome size variation among *F. deltoidea* variants can assist taxonomists to ascertain the taxonomic rank and avoid misidentification of variants. The genome size data of *F. deltoidea* is also required for genome sequencing projects as it will help reduce the cost and time needed. Up to now, the cytological data of *F. deltoidea* such as chromosome number is still lacking. It is recommended to conduct further cytogenetics studies such as chromosome number and physical mapping on this species. The genome size and chromosome number data of this species will contribute to the species classification as well as future genome sequencing projects.

REFERENCES

Abdulla, M.A., K.A.A. Ahmed, F.M. Abu-Luhoom & M. Muhanid (2010). Role of *Ficus deltoidea* extract in the enhancement of wound healing in experimental rats. *Biomedical Research*, 21: 241-245.

Abdullah, Z., Hussain, K., Zhari, I., Rasadah, M. A., Mazura, P., Jamaludin, F. & Sahdan, R. (2009). Evaluation of extracts of leaf of three *Ficus deltoidea* varieties for antioxidant activities and secondary metabolites. *Pharmacognosy Research* 1: 216-23.

Adam, Z., Ismail, A., Khamis, S., Mokhtar, M. H. M. & Hamid, M. (2011). Antihyperglycemic activity of *F. deltoidea* ethanolic extract in normal rats. *Sains Malaysiana*, 40(5): 489-495.

Adam, Z., Khamis, S., Ismail, A. & Hamid, M. (2012). *Ficus deltoidea*: A potential alternative medicine for diabetes mellitus. *Evidence-Based Complementary and Alternative Medicine*, 2012: 1-12.

Albach, D. C. & Greilhuber, J. (2004). Genome size variation and evolution in Veronica. *Annals of Botany*, 94(6): 897–911.

Bennett, M. D. & Leitch, I. J. (2005). Plant genome size research: A field in focus. *Annals of Botany*, 95(1): 1–6.

Brown, S. (1994). Applications of Flow Cytometry in plant biology and biotechnologies. *Review and Perspectives, Biotechnology & Biotechnological Equipment*, 8(3): 75-78.

Bunawan, H., Mat Amin, N., Bunawan, S. N., Baharum, S. N. & Mohd Noor, N. (2014). "*Ficus deltoidea* Jack: A Review on Its Phytochemical and Pharmacological Importance." *Evidence-Based Complementary and Alternative Medicine*, 2014: 1-8.

D'Hondt, L., Hofte, M., Bockstaele, E. V. & Leus, L. (2011). Applications of flow cytometry in plant pathology for genome size determination, detection and physiological status. *Molecular Plant Pathology*, 12(8): 815-828.

Doležel, J. & Bartoš, J. (2005). Plant DNA flow cytometry and estimation of nuclear genome size. *Annals of Botany*, 95(1): 99-110.

Doležel, J. & Greilhuber, J. (2010). Nuclear genome size: Are we getting closer? *Cytometry Part A*, 77(7) 635–642.

Doležel, J., Greilhuber, J. & Suda, J. (2007). Estimation of nuclear DNA content in plants using flow cytometry. *Nature protocols*, 2(9): 2233-2244.

Hakiman M. & Maziah M. (2009). Non enzymatic and enzymatic antioxidant activities in aqueous extract of different *Ficus deltoidea* accessions. *Journal of Medicinal Plants Research*, 3(3): 120–131.

Hasan, N. N. F., Nashriyah, M., Abdul R. N. Z., Mazlan N. Z., Haron, N, Mahmud K., Abdul, Y. G. & Abdul, M. A. (2012). Morphological phylogenetic analysis of seven varieties of *Ficus deltoidea* Jack from the Malay Peninsula of Malaysia. *Public Library of Science* 7(12): 1-7.

Hendrix, B. & Stewart, J. M. (2005). Estimation of the nuclear DNA content of Gossypium species. *Annals of Botany*, 95: 798-797.

Hojsgaard, D. & Herbig, E. (2013). Flow Cytometry applied to plant reproductive biology. *American laboratory*, 45(10): 8-13.

Jedrzejczyk, I. & Sliwinska, E. (2010). Leaves and seeds as materials for flow cytometric estimation of the genome size of 11 Rosaceae woody species containing DNA-staining inhibitors. *Journal of Botany*, 2010: 1–9.

Kochummen, K. M. & Rusea, G. (2000). Moraceae. *Tree Flora Sabah Sarawak*, 3: 181- 334.

Loureiro, J., Rodriguez, E., Doležel, J. & Santos, C. (2006). Comparison of four nuclear isolation buffers for plant DNA flow cytometry. *Annals of Botany*, 98: 679–689.

Madon, M., Phoon, L. Q., Clyde, M. M. & Mohd, D. A. (2008). Application of flow cytometry for estimation of nuclear DNA content in *Elaeis*. *Journal of Oil Palm Research*, 20: 447-452.

Midin, M. M., Nordin, M. S., Madon, M., Saleh, M. N., Goh, H-H. & Normah, M. N. (2017). Determination of the chromosome number and genome size of *Garcinia mangostana* L. via cytogenetics, flow cytometry and k-mer analyses. *Caryologia*, 71: 35-44.

Ochatt, S. J. (2008). Flow cytometry in plant breeding. *Journal of quantitative cell science*, 73: 581-598.

Pellicer, J., Hidalgo, O., Dodsworth, S., & Leitch, I. J. (2018). Genome size diversity and its impact on the evolution of land plants. *Genes*, 9(2): 1-14.

Ripain, I. H. A., Vikram, P., Kayat, F., Susanto, D. & Aurufullah, M. (2015). In vitro shoot regeneration of Mas Cotek (*Ficus deltoidea* Jack) A valuable Malay Medicinal Plant. *Journal of Tropical Resources and Sustainable Science*, 3 (2015): 46-50.

Roberts, A. V. (2007). The use of bead beating to prepare suspensions of nuclei for flow cytometry from fresh leaves, herbarium leaves, petals and pollen. *Cytometry*, 71(12) 1039-1044.

Rosnah, J., Khandaker, M. M. & Boyce, A. N. (2015). *Ficus deltoidea*: Review on Background and Recent Pharmacological Potential. *Journal of Agronomy*, 14(4): 310-318.

Subhash, J. B., Nurul, A. H. & Farida, H. S. (2009). Genetic variability based on random amplified polymorphic DNA in mistletoe fig (*Ficus deltoidea* Jack) collected from Peninsular Malaysia. *Journal of Forest and Environmental Science*, 25 (1): 57–65.

Sulaiman, M. R., Hussain, M. K., Zakaria, Z. A., Somchit, M. N., Moin, S., Mohamad, A. S. & Israf, D. A. (2008). Evaluation of the antinociceptive activity of *Ficus deltoidea* aqueous extract. *Fitoterapia*, 79(7): 557–561.

Vrána, J., Cápal, P., Bednářová, M. & Doležel, J. (2014). Flow Cytometry in Plant Research: A Success Story. In: Nick, P. & Opatrny, Z. (eds.). Applied Plant Cell Biology, Plant Cell Monographs 22. pp. 395–430. Berlin: Springer.

Zakaria, Z. A., Hussain, M. K. Mohamad, A. S., Abdullah, F. C., & Sulaiman, M. R. (2012). Anti-inflammatory activity of aqueous extract of *Ficus deltoidea*. *Biological Research for Nursing*, 14: 90-97.

Adventitious Rooting of Fig (*Ficus carica* L.) Stem Cuttings in The Sandy Bris Soil

Salleh, M.S.[1*], Nordin, M.S[1]., Shahari, R.[1], Tajudin, N.S.[1] and Che Amri, C.N.A.[1]

[1]*Department of Plant Science, Kuliyyah of Science, International Islamic University Malaysia (IIUM), 25200 Kuantan, Pahang, Malaysia.*
Corresponding author: msyahmi@iium.edu.my

ABSTRACT

The most suitable type of fig (*Ficus carica* L.) to be grown in Malaysia is common fig. However, the seeds of common fig are non-viable. Thus, the planting material is usually produced through stem cuttings by using peat moss as propagation media. Although an acceptable rate of adventitious rooting could be achieved, the cost of production is substantial. Substituting peatmoss with under utilized sandy beach ridges interspersed with swales (BRIS) soil which contain more than 90% of sand may significantly reduce the cost of production. Hence, this research was conducted to investigate the potential of substituting peatmoss with sandy BRIS soil as propagation media for adventitious rooting of fig stem cuttings. The experiment was conducted using factorial randomized complete block design (RCBD). Two types of stem cutting mainly semi-hardwood (C1) and softwood (C2) were used as propagation material. Propagation media used were the mixture of peatmoss (PM) and BRIS soil (BS) in the following ratios of PM:BS mainly 100% PM (M1), 75:25 (M2), 50:50 (M3), 25:75 (M4), and 100% BS (M5), respectively. Survival rate and stem anatomy were recorded and observed at the end of the experiment (day 56 after sowing). Stem anatomy of C1 was slightly different compared to C2 with the presence of cork and cork cambium layers. The C1 recorded 100% survival rate under M2 as compared to C2 of only 80%. However, under M5, the C2 recorded significantly higher survival rate at 60% as compared to C1 at 0%. Hence, the present study had highlighted the potential of substituting peat moss with BRIS soil for the propagation of fig stem cutting.

Keywords: *Ficus carica* L., cuttings propagation, adventitious rooting, sandy BRIS soil.

INTRODUCTION

The fig (*Ficus carica* L.) is known to have high nutritional and medicinal values including antioxidant, anti-cancer and anti-bacterial properties (Mawa et al., 2013). Fig contains about 26.02 ± 0.63 g carbohydrate, 10.67 ± 0.31 mg vitamin C, 266.67 ± 2.78 mg potassium, 125.44 ± 3.37 mg calcium per 100 g of fresh fruit and 1.11 mg phenolic, 83.82 µg flavonoid, 41.72 µg anthocyanin, 2.06 µg tannin, and 29.51 mg mL^{-1} of antioxidant activities per gram of fresh fruit (Mahmoudi et al., 2018). According to Manago (2006), the fig is native to western Asia and has been cultivated for thousands of years in the Mediterranean countries of Europe and North Africa. Although this plant needs a warm and dry temperature to grown, it may adapt to several climates and could be cultivated in both sub-tropical and temperate regions.

Fig was introduced into Malaysia in the late 2000 and since then more than 800 varieties had been brought in by several growers. Recently, fig planting is becoming popular in Southeast Asia such as Thailand, Indonesia and Malaysia with growing interest among locals to grow their own fig tree (Syufihuddin et al., 2021). In fact, there is also increasing market demand on fresh fruit and its downstream products such as tea, ice cream and chocolates. In the year 2016 alone, around 308,460 ha of area in the world was harvested with fresh fig yield of 34,055 hg/ha that has world production of 1,050,459 tonnes with the top three producers are Turkey, Egypt and Morocco (FAO, 2017).

Hence, the fig has a great potential to become one of the alternative crops in the tropics. Based on the experience of local fig grower, fig trees could be successfully grown under rain shelter structure compared to open field planting due to high humidity and rainfall amount in Malaysia. Under tropical conditions, fig trees may produce fresh fruits all year round as compared to the seasonal nature of temperate climatic conditions. However, the fig planted in Malaysia is of the type known as the 'common type', which has the ability to produce fleshy fruits but the seeds are small and non-viable. Thus, the fig is usually propagated using vegetative propagation mainly stem cutting due to the parthenocarpic characteristics of the fig fruit which produces non-viable seeds; hence delaying domestication of this fruit in Malaysia due to the scarcity of planting materials. According to Stover et al., (2007), stem cutting is the most common method used worldwide for fig propagation.

In Malaysia, the planting material is usually produced through stem cuttings by using peat moss as propagation media. Although an acceptable rate of adventitious rooting could be achieved, the cost of production is substantial. Substituting peat moss with underutilized sandy beach ridges interspersed with swales (BRIS) soil which contain more than 90% of sand may significantly reduce the cost of production. According to Hanafi et al. (2010), the sandy Spodosols (BRIS soils) in Malaysia are not well utilized for crop production due to their inherent poor fertility, low water holding capacity, low organic content, and high infiltration rate. Hence, this research was conducted to explore the potential of substituting peat moss with sandy BRIS soil as propagation media for adventitious rooting of fig stem cuttings. In addition, cross section of stem cuttings were also observed at the end of the experiment to determine presence of root primordia cells.

MATERIALS AND METHODS

Experimental Design
The experiment was conducted at the Glasshouse and Nursery Complex (GNC), Kulliyyah of Science, International Islamic University Malaysia (IIUM) Kuantan using factorial randomized complete block design (RCBD) with five replications as shown in Figure 1. Two types of stem cutting mainly semi-hardwood (C1) and softwood (C2) were used as propagation material. These type of stem cutting were chosen due to their availability and abundance in the mother plant. The length and diameter of the stem cuttings were standardized at about 12.0 cm (±1.0 cm) in length and 0.7 cm (±1.0 cm) as shown in Figure 2, respectively. All cuttings were dipped in powder rooting hormone before sowing. Propagation media used were the mixture of peatmoss (PM) and BRIS soil (BS) in the following ratios of PM:BS mainly 100% PM (M1), 75:25 (M2), 50:50 (M3), 25:75 (M4), and 100% BS (M5) as shown in Figure 3. The pH of propagation media were also recorded prior to the experiment.

Fig. 1: Experimental Units Arranged in the RCBD.

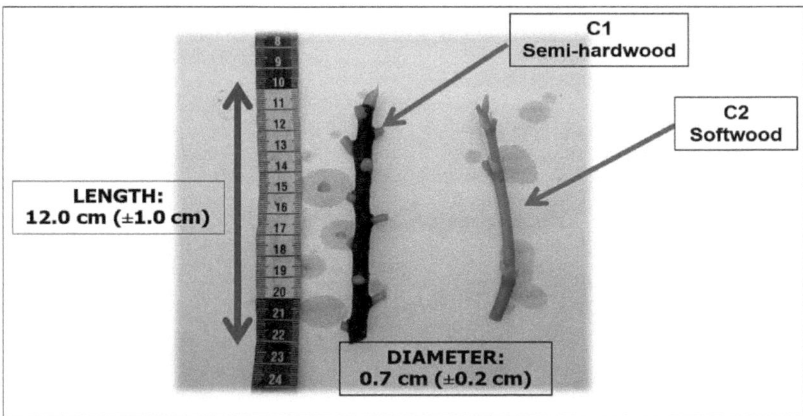

Fig. 2: Types, Length, and Diameter of Stem Cuttings.

Fig. 3: The Physical Appearance and pH of the Propagation Medium.

Data Collection and Statistical Analysis

The survival rate of stem cuttings in different types of propagation medium were recorded on day 56 of the experiment by using the following formula:

$$\text{Survival rate (\%)} = \frac{\text{No. of survived cuttings} - \text{No. of died cuttings}}{\text{Total no. of cuttings}} \times 100\%$$

In addition, the stem anatomy of cuttings were also observed at the end of the experiment following the steps shown in Figure 4. In brief, survived cuttings were collected and cut into smaller pieces of between 30 – 50 mm long and fixed in the formalin aceto-alcohol (FAA) solution. Fixed samples were then rinsed and processed in an automated Thermo Scientific™ ExcelsiorAS™ Tissue Processor (Thermo Fisher Scientific, USA). The stem cuttings were then embedded into a paraffin wax block using the Thermo Scientific™ HistoStar™ Embedding Workstation (Thermo Fisher Scientific, USA). Embedded tissues were then sectioned using the rotary microtome (Thermo Fisher Scientific, USA) at 8 µm thickness and transferred onto glass slides. All sections were then examined and photos recorded using light microscope with attached camera (Figure 4). All collected data were analysed using the analysis of variance (ANOVA) followed by the mean comparison analysis using Duncan's multiple range tests (DMRT). The statistical analysis was carried out using the Statistical Analysis System (SAS) version 9.1.

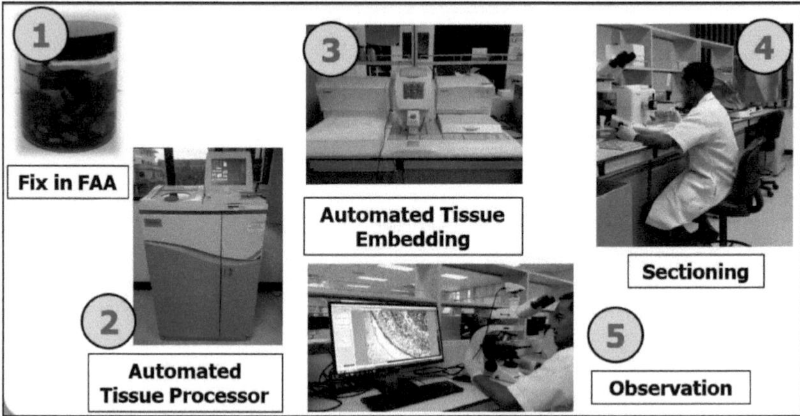

Fig. 4: The Procedures for Stem Anatomy Analysis.

RESULTS AND DISCUSSION

Cutting is the propagation method in which the induction of adventitious rooting under suitable conditions would give rise to a seedling (Hartmann et al., 2011). According to Boliani et al., (2019), cutting is the most common propagation method used for cloning of woody plants on a large scale. However, Han et al., (2009) highlighted that types of cuttings and propagation medium may directly influence the rooting of fig cuttings. As shown in Figure 5, the highest survival rate was recorded by the C1 (semi-hardwood) cutting in M2 (75% PM : 25% BS). In contrast, C2 only recorded 80% survival rate under the M2 medium. However, the survival rate of C1 cutting decreases with increasing substitution of peatmoss with BRIS soil (Figure 5). In fact, under 100% BRIS soil (M5), none of the C1 cuttings survived. The C2 cutting on the other hand recorded 80% and 60% survival rate under M4 (25% PM : 75% BS) and M5 (100% BS), respectively.

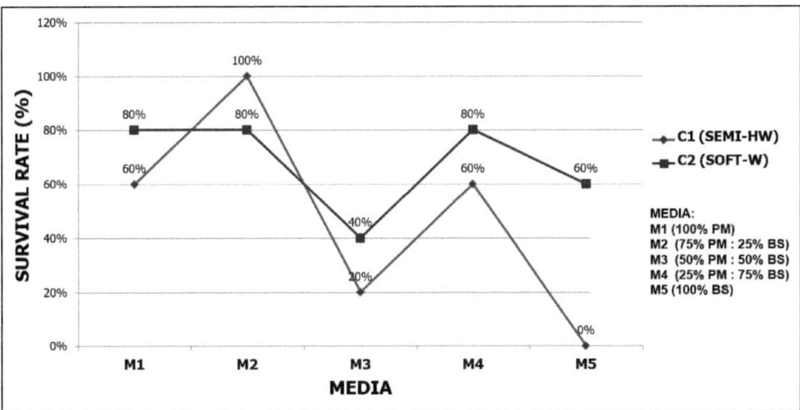

Fig. 5: Survival Rate of Stem Cuttings at the End of the Experiment.

Propagation substrate is the medium where roots develop, thus the material must have good physicochemical characteristics, free from soil pathogens, and high economic viability (Boliani et al., 2019). The optimum propagation medium however was found to be species and types of cutting's specific (Han et al., 2009). Sand may provide good aeration for adventitious rooting but lack in nutrients as compared to peat moss (Hartmann et al., 2011). In addition, peat moss may not only supply sufficient amounts of nutrients and organic matter, but also moisture for adventitious rooting of cutting (Wahome et al., 2011).

However, excessive supply of moisture due to overwatering of peat moss may promote disease and fungal growth on the cuttings (Hartmann et al., 2011). Waseem et al. (2013) previously highlighted that an effective propagation medium should be able to hold sufficient amount of moisture and nutrients, while at the same time provide aeration and porosity for root growth and development. Hassanein, (2013) also emphasized that the propagation medium should hold sufficient moisture to promote faster rooting of cutting. Indeed, if the rate of transpiration is higher than the rate of water absorption, the stem cuttings might wilt and eventually die (Okunlola and Akinpetide 2016).

Hence, a combination of 75% peatmoss and 25% BRIS soil in M2 might be perceived as the most suitable propagation medium of fig stem cuttings due to higher survival rate at 100% for C1 and 80% for C2 (Figure 5). This probably due to the ability of M2 medium to supply sufficient amount of moisture and nutrients while at the same time provide sufficient aeration and porosity for adventitious rooting of the cuttings. Previous study by Syufihuddin et al. (2021) also indicated that the highest survival rate of the hardwood fig stem cutting was recorded in the combination of peatmoss:perlite at 1:1 ratio used as the propagation medium. The stem anatomy of softwood cutting was slightly different with the semi-hardwood with the absence of cork and cork cambium layers (Figure 6). In the softwood cutting, the cork and cork cambium layers were still not developed. However, the cortex layer which is regarded as the origin of cork and cork cambium layers could be clearly observed in the softwood cutting as shown in Figure 6.

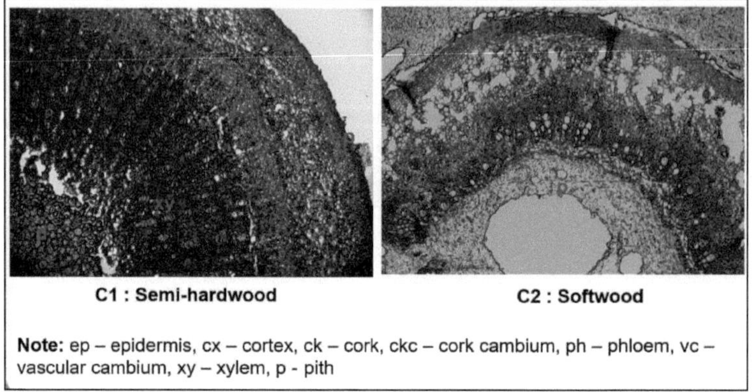

Note: ep – epidermis, cx – cortex, ck – cork, ckc – cork cambium, ph – phloem, vc – vascular cambium, xy – xylem, p - pith

Fig. 6: The Anatomy of Stem Cuttings Prior to the Experiment.

Moreover, as mentioned by Hartmann et al., (2011), the process of adventitious rooting in stem cuttings was initiated with the dedifferentiation of vascular cambium cells mainly xylem and phloem into root primordia cells (Figure 7). Optimum condition of propagation medium would further enhance the process of adventitious rooting Boliani et al., (2019). This will led to the protrusion of root primordia cells from the epidermis layer of the stem (Figure 8). Syufihuddin et al. (2021) stated that the propagation by stem cutting involves the initial root formation of the stem. The initial root formation of the stem is also known as the adventitious rooting of stem cutting (Pio et al., 2005; Hartmann et al., 2011). Root plays a vital role in plants, especially anchorages, as well as in the transport of nutrients and water throughout cutting during early growth, hence proper media selection is, therefore, needed to enhance healthy root development (Syufihuddin et al., 2021). Previously, Bisi et al. (2016) observed that the rooting percentage of the herbaceous cuttings of Lemon, 'Bêbara Branca' and 'Pingo de Mel' fig cultivars were superior than the woody cuttings. In addition, Pio et al. (2004) also reported that the herbaceous cuttings of ' Roxo de Valinhos' fig cultivar showed higher rooting percentages and higher number of shoots in the presence of leaves. However, in the present study, both types of cutting may produce an adventitious root as reflected by the survival rate of cuttings. However, types of propagation medium would influence the process of adventitious rooting thus affect the survival rate of the cuttings.

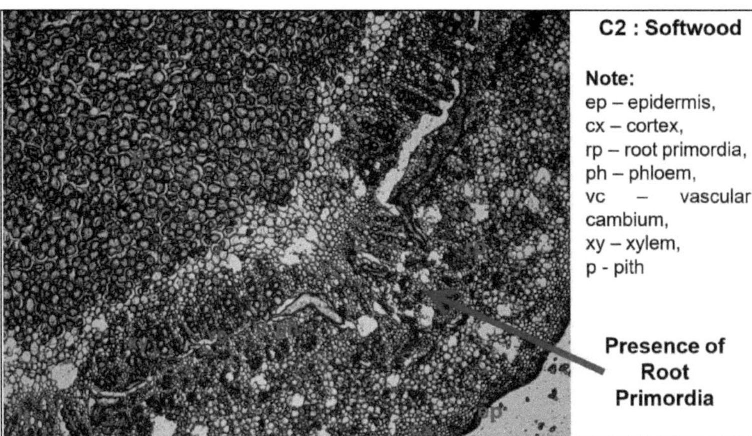

Fig. 7: Anatomy of Softwood Stem Cutting at the End of Experiment.

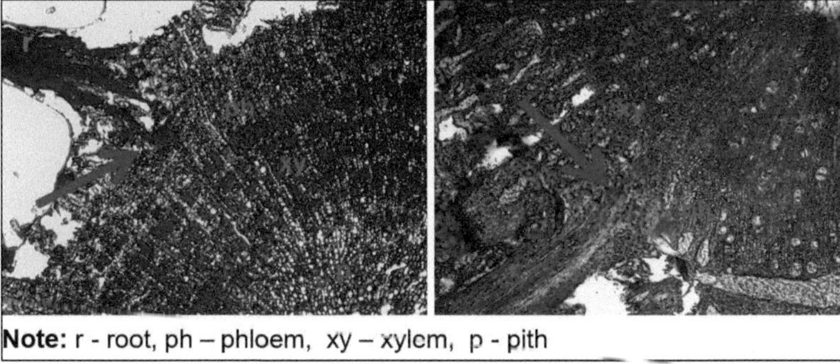

Fig. 8: Anatomy of Semihardwood Stem Cutting at the End of Experiment.

CONCLUSION

The stem anatomy of C1 (semi-hardwood) was slightly different from C2 (softwood) due to the presence of the cork and cork cambium layers. The C1 recorded 100% survival rate under M2 (75% peat moss : 25% BRIS soil) as compared to C2 of only 80%. However, under M5 (100% BRIS soil), the C2 cutting recorded a significantly higher survival rate at 60% as compared to C1 at 0%. This indicated that 100% BRIS soil may be used in fig propagation using softwood stem cutting as propagation material. However, if semi-hardwood cutting is being used, the advisable rate of peat moss substitution with BRIS soil was about 25% (M2). In conclusion, the present study had highlighted the potential of substituting peatmoss with BRIS soil for the propagation of the fig stem cutting.

ACKNOWLEDGEMENT

We would like to thank the Ministry of Higher Education Malaysia for the financial support to the International Islamic University Malaysia (IIUM) through the Research Initiative Grant Scheme (RIGS), IIUM (RIGS15-128-0128), and the Kuliyyah of Science, IIUM Kuantan for providing research facilities.

REFERENCES

Bisi, R.B., Locatelli, G., Barbosa, C.M.A., Pio, R., and Balbi, R.V. (2016). Rooting of stem segments from fig tree cultivars. Acta Scientiarum Agronomy, 38(3): 379-385.

Boliani, A.C., Ferreira, A.F.A., Monteiro, L.N.H., daSilva, M.S.C., and Rombola, A.D. (2019) Advances in propagation of *Ficus carica* L., Revista Brasileira de Fruticultura, 41(3): 1-13.

FAO. (2017). Figs world statistical data. *FAOSTAT.* Retrieved April 17, 2021, from http://www.fao.org/faostat/en/#data/QC

Han, H., Zhang, S., and Sun, X. (2009). A review on the molecular mechanism of plants rooting modulated by auxin. African Journal of Biotechnology, 8(3): 348-353.

Hanafi, M.M., Shahidullah, S.M., Niazuddin, M., AbdAziz, Z., and Mohammud, C.H. (2010). Potential use of sea water for pineapple production in BRIS soil. *International Journal of Agricultural Biology,* 12: 396-400.

Hartmann, H.T., Kester, D.E., Davis Jr., F.T. and Geneve, R.L. (2011). Hartmann and Kester's plant propagation (8th Edition). Upper Saddle River, NJ: Prentice Hall.

Hassanein, A.M.A. (2013). Factors influencing plant propagation efficiency via stem cuttings. *Journal of Horticultural Science & Ornamental Plants,* 5(3): 171–176.

Hitchcock, A. E. (1928). Effect of Peat Moss and Sand on Rooting Response of Cuttings. *Botanical Gazette,* 86(2):121-148.

Mahmoudi, S., Khali, M., Benkhaled, A., Boucetta, I., Dahmani, Y., Attallah, Z. and Belbraouet, S. (2018). Fresh figs (*Ficus carica* L.): Pomological characteristics, nutritional value, and phytochemical properties. European Journal of Horticultural Science. 83(2), 104-113.

Manago, N. (2006). Fig. In The Japanese Society for Horticultural Science (eds.). *Horticulture in Japan 2006.* Shoukadoh Publication, Dept. of Publishing of Nakanishi Printing Co., Ltd. pp. 106 – 110.

Mawa, S., Husain, K., and Jantan, I. (2013). Ficus carica L. (Moraceae): Phytochemistry, Traditional Uses and Biological Activities. *Evidence-Based Complementary and Alternative Medicine,* 2013: 1-8.

Okunlola, A.I. and Akinpetide, E.O. (2016). Propagation of Ficus benjamina and Bougainvillea spectabilis using different media. *Advances in Agriculture and Agricultural Sciences,* 2(2): 21–27.

Pio, R., Araújo, J.P.C., Bastos, D.C., Alves, A.S.R., Entelmann, F.A., Scarpare, F.J.A., Mourão, F.A.A. (2005). Substrates in the rooting of fig tree herbaceous cuttings originated from the sprouting. Ciência e Agrotecnologia, 29(3): 604-609.

Pio, R., Chalfun, N.N.J., Ramos, J.D., Gontijo, T.C.A, Toledo, M., Carrijo, E.P. (2004). Presence of leaves and apical bud in the rooting of fig tree herbaceous cuttings originating from sprouting. Revista Brasileira Agrociência, 10(1): 51-54.

Stover, E., Aradhya, M., Ferguson, L., and Crisosto, C.H. (2007). The Fig: Overview of an Ancient Fruit. *Hortscience,* 42(5): 1083-1087.

Syufihuddin, M.S., Rozilawati, S., Che Nurul Aini, C.A., Nur Shuhada, T., Razali, M.M. and Salleh, M.S. (2021). Early development of fig (*Ficus carica* L.) root and shoot using different propagation medium and cutting types. *Tropical Life Sciences Research,* 32(1): 83–90. https://doi.org/10.21315/tlsr2021.32.1.5

Wahome, P.K., Oseni, T.O., Masarirambi, M.T. and Shongwe, V.D. (2011). Effects of different hydroponics systems and growing media on the vegetative growth, yield and cut flower quality of gypsophila (*Gypsophila paniculata* L.). *World Journal of Agricultural Sciences,* 7(6): 692–698.

Waseem, K., Hameed, A., Jilani, M.S., Kiran, M., Javeria, S. and Jilani, T.A. (2013). Effect of different growing media on the growth and flowering of stock (Matthiola incana) under the agro-climatic condition of Dera Ismail Khan. *Pakistan Journal of Agricultural Sciences,* 50(3): 523–527.

Pests and Diseases of *Ficus* sp. In Malaysia

Azmi, N.S.A[1], Nurul Huda, A.[1*],Nurul Wahida, R.[2*]
[1]Department of Plant Science, Kulliyyah of Science, International Islamic University Malaysia (IIUM), Kuantan, 25200, Pahang, Malaysia.
[2]Faculty of Plantation and Agrotechnology, Universiti Teknologi MARA (UiTM), Cawangan Melaka Jasin Campus, 77300 Merlimau, Melaka, Malaysia.
Corresponding author email: anhuda@iium.edu.my, nurulwahida87@uitm.edu.my

ABSTRACT

Nowadays, the value of fig cultivation in Malaysia increases as many species of *Ficus* grown as an ornamental in urban landscapes and popular indoor plants for the hobbyist or as a commercial production for fruit and planting material (cuttings, bareroots etc). However, *Ficus* plants are susceptible to diseases and pests such as anthracnose, *Botrytis* blight, gray mold, fig rust, *Cercospora* leaf spot and fruit rot. Furthermore, they are also prone to the attack by xylaphagous insects, sap sucking insects and plant eating insects. Therefore, maintaining healthy *Ficus* flora becomes a great challenge especially from sudden outbreaks of pests and pathogens. Hence, this chapter will describe the major pests and diseases in fig cultivation in Malaysia.

Keyword: Xylophagous. Sap sucking. Plant eating. Anthracnose. *Cercospora. Botrytis. Cerotelium fici. Alternaria alternate*

INTRODUCTION

Ficus sp. or well known as fig play a major role as an important fruit crop especially in tropical and subtropical countries. Fig is one of the oldest cultivated plants and originates in the Mediterranean region, with Turkey becoming one of the largest producers of figs, which cover 30% of the whole world production (Schubert et al., 1999). *F. carica* cultivation in Malaysia can produce fruit throughout the year due to advanced technology and suitable climate even though this fig is not indigenous in South East Asia but native to the Mediterranean and Western Asia. Basically, figs have been distributed throughout the world including European countries, New World countries and Asian (Flaishman et al., 2008) and it has been harvested worldwide for consumption in the form of dry or fresh. The fruits contain an important and high nutritional value such as a source of vitamins, minerals as well as high amounts of fiber and polyphenol. Other than that, fig has a beneficial role in therapeutic remedy as well as medicinal purpose to treat certain diseases (Mawa et al., 2013).

In Malaysia, planting fig trees is still in the early footstep phase, but it has been increasing as Malaysian started to grow fig trees for hobby activities and commercialization. Fig has been cultivated widely, and the plantations can be found in many states across Malaysia, with Perlis becoming one of the largest producers in Malaysia (Kamarubahrin et al., 2019). Fig varieties that are well-suited to the Malaysia environment are such as Black Jack, Brunswick and Texas Everbearing. However, figs that are commonly grown under an open field system easily get attacked with diseases and pests. There are several diseases and pests which commonly attack *Ficus* sp. In this chapter, we will briefly discuss the pests and diseases that occur in fig cultivation.

PEST OF *Ficus*

There is no particular pest that has occurred in the cultivation of *F. carica* under greenhouse conditions (control environment, chemical pesticide-free with strict and sustainable farming practices) based on the informal surveys on few commercial figs (*F. carica*) producers. Recent studies suggested that not all varieties of *Ficus* trees or their relatives are susceptible to insect pest infestations (Bragard et al., 2021).

However, some pests are associated with this plant genus (i.e *Ficus*) in agriculture and in nature. Some *Ficus* species that are noted to be resistant to whitefly infestation are *F. microcarpa* "Green Island", *F. religiosa*, *F.carica*, *F. lyrata*, *F. pumila* and *F.elastica* "Burgundy", whereas *F. benjamina* and *F. nitida* are susceptible (Nair et al., 2015). *Ficusbenjamina*, however, is susceptible to infestation of leaf feeding caterpillars like *Triloca varians* (Bombycidae) (Basari et al., 2019). Pest of fig can be divided into a few groups which are xylophagous, sap-sucking, plant-eating (including fruit-, leaf- and root-eater) insects and mites. A few species representing the groups were described in this chapter.

Xylophagous
Xylophagous insects refer to certain insects or insect larvae that feed on or live-in wood. These insects develop either in living woody plants or dying trees and shrubs or dead parts of still living woody plants. Some species may feed on a minor part of the living plant and do not cause any significant harm but still possible to have a serious impact on the health of the plant if the population is massively abundant. Mortality also can occasionally occur especially on susceptible plant species. Xylophagous may become a vector for wood rutting or staining fungi (Scudder & Cannings, 2005) or other secondary attacks which further disrupt the plant health. These pathogens often have far more serious impacts on the plant's health than the insect's feeding itself.

Most of the xylophagous species come from Order Coleoptera of the families Bostrychidae (auger beetle), Buprestidae (jewel beetle), Cerambycidae (long-horned beetle) and Curculionidae (weevil). *Hypoborus ficus* Erichson (Scolytinae: Curculionidae), *Euwallacea fornicates* (Scolytinae: Curculionidae), *Hypothenemus leprieuri* Perris (Scolytinae: Curculionidae), *Hesperophanes griseus* Fabricius (Cerambycidae), *Scobicia chevrieri* Villa (Bostrichidae) and *Sinoxylon sexdentatus* Oliver(Bostrichidae) are among xylophagous insect that reported damaging fig trees by many literatures.Three of the species mentioned, which were *Hypoborus ficus*, *Hesperophanes griseus*, *Scobicia chevrieri* and *Sinoxylon sexdentatus* commonly reported causing losses by damaging both stems and branches of *Ficus* tree in Turkey (Aksit et al., 2005). The larva makes tunnels in phloem and xylem tissues and under the bark. The adults mostly feed on flowers and foliage of some plants, except for Scolytids and Bostrychids, which feed in wood tissue. All the species usually preferred branches that were weakened or dried up for other reasons.

a) *Hypoborus ficus* Erichson
A very small Scolytids beetle from the family Curculionidae. The adult beetle has a stout body with a length around 1-1.4 mm in females, while the male is smaller. The body is brown in color and covered with plumose, scale-like whitish setae. Heads are rounded and partly concealed below the pronotum from the dorsal view. The pronotum is wider than long and punctate. Fore wings (elytra) cover the posterior end of the abdomen.

Fig bark beetles usually attack weakened trees and hasten the withering process. This insect can cause the death of branches and trees if present in a high number of individuals. Adult beetles construct an entrance tunnel through the bark. A nuptial chamber was then excavated at the end of the entrance tunnel with one or more egg tunnels extended from it. Hatched larvae were then tunneled perpendicularly deeper into the trunk, which led to increases of larvae galleries filled with excrement as the larvae grew. Infestations occur throughout the entire year with three to more generations annually (Alford, 2014). Natural enemies such as *Pyemotes johnmoseri* that was first documented as a parasitoid of this small bark-burrowing beetle in Ukraine (Khaustov, 2004). Aksit et al. (2007) also reported that *P. johnmoseri* parasitizes mostly on larvae (90.5%) and occasionally on eggs or pupae of *H. ficus*.

Sap-sucking
Sap-sucking insects, on the other hand, eat or consume the sap of a plant. The sap is the liquid that maintains hydration and transports vital nutrients throughout the plant's body. Sap-sucking insects can severely detriment a plant by reducing its vigour but uncommonly caused mortality to the plant. Examples of sap-

sucking insects include scale insects, mealybugs, aphids, whiteflies and thrips. Mealybugs (Hemiptera: Pseudococcidae) are one of the most serious insect pests affecting fig production (Zada et al., 2003). Three species of mealybugs; (*Phenacoccus solenopsis*, *Pseudococcus longispinus* and *Planococcus ficus* were reported from experimental fig orchards in Malaysia (Moniruzzaman et al., 2017). Infestation at the final stage of fruit development or prolonged infestation by mealybug can cause fruit spoilage.

Plants can and most often do recover from an infestation of sap-sucking insects but since they also can transfer plant pathogens like bacteria and viruses thus infected plants may die from the second issue related to feeding activity. When hemipterans infest fruit or excrete honeydew that covers fruit and leaves, often resulting in sooty mold growth. The sooty mold growing on these sweet secretions densely covers the plant surface and restricts the light absorption by the leaves resulting in defoliation and loss of height increment. Ants are often attracted to feed on the honeydew excreted by the sap-sucking insects (**Figure 1**).

Fig. 1: Sugary poop excretes by scale insects attract ants to the infested plant.

a) *Planococcus ficus* Signoret
This insect, known as vine mealybugs, is primarily found in countries of tropical and subtropical regions. This insect lives in colonies on different types of plants (figs, apples, grapes, citrus, dates, mango, etc.) mainly for their fruits. The shape and size between sexes are almost similar during nymphal stages but different in adult form with females are much larger than males usually with an approximate length of 4 mm and oval body covered with fine white powdery wax, wingless and have the waxy hair-like extension. Meanwhile, the adult male is very minutes and fragile with less than 1 mm long and equipped with a pair of inconspicuous transparent wings, beaded antennae, and two long filaments at the end of the abdomen (anal seta), which used for flight stabilization and have reduced mouthparts.

Overfeeding by extremely large colonies will weaken the plant and sometimes kill the plant but most of the time it is just fruit or infested plant parts that are destroyed (Berning et al., 2014). Meanwhile, a sooty mold grows on their waxy secretions, reducing the quality of the fruits and sometimes making them inedible or unmarketable. Predatory insects, particularly lady beetles (Coccinellidae), are a natural biocontrol for mealybugs infestations.

b)*Trioza brevigenae* Mathur

Ficus leaf-rolling psyllid is a new pest reported infestedChinese Banyan tree *Ficus microcarpa*. In contrast with fig psylla, *Homotoma ficus* causes damage like other hemipterans by sucking plant sap and giving rise to sooty mold problem (Gencer et al., 2007). *T. brevigenae* is associated with rolled leaves problem. Adult psyllid is small, about 2.6-2.8 mm long and usually found outside and adjacent to the rolled leaves. The anterior part (head and thorax) is brownish-green, while the abdomen green but becomes darker when mature. This insect has protruding red eyes, and its wings are transparent with no pattern.

Infestation of this hemipteran on *Ficus microcarpa* shows symptoms where young leaves at the branch and twig tips are completely and tightly rolled into a narrow cylinder. The rolling process begins at the leaf apex, progresses adaxially (upper leaf surface) along each margin, thus forming two waves. One rolled margin will eventually overtake the other, creating a cylinder with two tubes. The rolled leaf is home to various nymphal stages of *T. brevigenae* and is brittle if peeling back but remains green in color if there is no secondary attack by other pests (Hodel et al., 2016).

Usually, psyllids tend to be very host-specific. Compared to fig psylla *Homotoma ficus* andAsian citrus psyllid *Diaphorina citri*, information about ficus leaf-rolling psyllid and their management is still lacking. However, frequent monitoring and constantly removing infested leaves by periodic pruning might be effective management to control this pest.

c) *Aleurodicus disperses* Russell

Spiraling whitefly *Aleurodicus* genus is from family Aleyrodidae has relatively large size compared to ficus whitefly *Singhiella simplex* and known to arrange their eggs in a spiral pattern interspersed with bits of the white wax underside of leaves (**Figure 2**). The eggs will hatch after a few days and undergo four nymph stages before emerging as adults. Nymphs flat dorsally exude tufts of white wax on the dorsal surface and additionally produce glass-like waxy rods on their lateral surfaces. The fourth stage is usually covered fully with wax before it starts to pupate. Mature pupae appear fluffy as they flocculate with copious amounts of white cottony wax. Adult spiraling whitefly is small, white, moth-like insects with 2–3 mm long. They are polyphagous and known to attack many plants including fruit trees, ornamentals, shade trees and weeds.

Adults and the first three nymphal stages feed by sucking plant sap which weakens the plant and usually causes leaves to turn yellow and fall prematurely (David & Regu, 1995). Like other hemipterans, spiraling whiteley also acts as a vector for plant viruses and extensive feeding activity often causes sooty mold (**Figure 3**). Most whitefly species can be controlled using chemical insecticide either soil or trunk application for long-term control. Meanwhile, foliar insecticides are usually used as an active control. In nature, this insect pest has natural enemies like parasitic wasp *Encarsia* spp. that lay eggs on immature stages and predatory insects such as lacewing, larvae of hoverfly and ladybird beetles (either adult or larvae) that attack all stages.

Fig 2: Spiral pattern of white wax made by *Aleurodicus dispersus*.

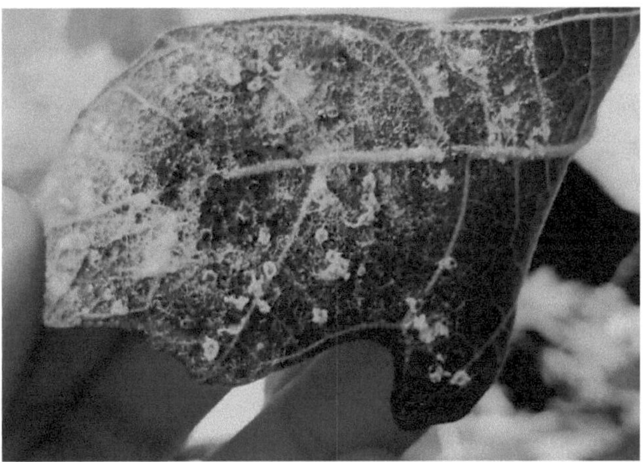

Fig. 3: Whitefly excrete honeydew and wax that covers leaves, often resulting in sooty mold growth especially in areas with high feeding activity.

d) *Singhiella simplex* Singh

Ficus whitefly nymphs suck plant juices and cause leaf chlorosis, defoliation and dieback. Heavy infestations of the pest caused clouds of adults and dripping honeydew and sooty mold. However, this pest does not produce the abundant white, waxy, flocculent material that is typically produced by other species. Adult females deposit the eggs along the midnerve on the abaxial leaf surface (Hodel, 2017). Eggs of *Ficus* whitefly are small, elongate, and yellow to light brown in colour. Ficus whitefly nymphs are light green to tan, flat, oval, semi-transparent and camouflaged with the surrounding leaf surface. The adult is 1.4-1.6 mm long (after one month) with typical white waxy wings and has grayish-brown markings (Hodel, 2017).

e) *Scirtothrips dorsalis* Hood

Thrips are a widespread pest that gradually achieves global distribution since it is an opportunistic generalist species that feeds on various plants from herbs to large trees (Kumar et al., 2009). This insect has a small body size (< 2 mm) and it is difficult to detect in fresh vegetation due to its rapid movement and cryptic nature. The female laid very tiny eggs that are inserted into the meristem or tender tissues of the host plant, and the time for egg development may last 7 days. Unlike other thrips species that pupate on soil, all *Scirtothrips dorsalis* life stages occur above-ground plant parts of its hosts with pupation taking place on the leaf axils.

Scarring damage due to feeding activity can be seen clearly when the affected plant parts mature and usually become an entry for the transmission of viruses and other pathogens (Riley et al., 2011). Feeding activity of adults and nymphs on leaves will cause the leaf to curl upward (leaf curl disease), shrivel of young leaf and reduction of leaf size. The sustainable management practices to control thrips infestation include crop rotation, removal of alternate hosts and supporting the maximum use of natural enemies including predators and parasites, and rotating insecticides to avoid resistance. Natural enemies include minute pirate bugs, *Orius* spp. (Hemiptera: Anthocoridae), predatory mite *Amblyseius* spp. and other predatory arthropods have been reported to effectively control thrip population in the field (Dogramaci et al., 2011).

f) *Gynaikothrips uzeli* Zimmermann

The feeding damage of *G. uzeli* appears sunken, reddish spots on leaves and the presence of folded-leaf galls. Severe damage leads to defoliation or stunted growth. Because Ficus spp. are primarily used as ornamental plants, the damage is more of an aesthetic value. Adult thrips use their sucking mouthparts to feed on new leaves, causing them to distort and deform. The leaf folds along the main vein and forms a permanent gall with dark brown, reddish, or purplish scarring. Prolonged feeding by high populations of adults and larvae can slow plant growth and leaves eventually drop from the plant prematurely. Infested trees will not be killed, but the ornamental value of the plant is reduced markedly (Funderbark et al., 2014).

G. uzeli is a dark-colored thrip with a stout body. Adults are 2.5-3.6 mm in length (average length of females and males is 2.8 mm and 2.2 mm respectively) with a brownish-black body. Both sexes possess whitish or pale fringed wings. Tarsi and most of the antennal segments are yellow. The head is elongated and slightly constricted behind the eyes. The duration of the life cycle depends on temperature, but under greenhouse conditions, it is completed in 30 days (Arthurs et al., 2011).

Plant-eating

Fig, like any other plant from the family Moraceae also faces problems caused by plant-eating insects. This group includes insect pests that feed on leaves, flowers, roots, fruits as well as seeds. **Figure 4** shows common symptoms caused by leaf-feeding beetles on *Ficus* leaf. Common insects from this group include several species of bagworm, caterpillars, and beetles, particularly chrysomelids and curculionids. Besides, adults of some xylophagous species are also included in this group, although the damage caused is less severe. The level of pest infestation by leaf-eating insects mostly influences factors such as tree density, leaf palatability, geographical distribution, and altitude (Basset & Novotny, 1999).

Fig. 4: Feeding damage by leaf beetle.

a) *Trilocha varians* Walker
This moth from the family Bombycidae is a leaf-defoliator to many plant species. The adult is medium size with a wingspan of 25–27 mm, dark brown to greyish brown in color. Their caterpillars have an elongated body, grey to dull brown with dark dots and a lateral row of black dots. A slight protuberance can be seen on each of the thoracic, 5th and 8th somites. There is also characteristic like a long slender horn on anal somite. Mature (2nd-4th instar) larvae are usually covered with white powdery secretions. Pupation takes place on a host plant.

This insect has various range of host plants like *Artocarpus communis, A. kamansi, A. heterpphyllus, Ficus benghalensis, F. benjamina, F. carica, F. elastica, F. infectoria, F. nitida* and *F. religiosa* (Arya, 2020). The destructive stage is the larvae and heavy feeding can cause complete defoliation (Navasero & Navasero, 2014) which disrupts photosynthesis and degrades the aesthetic value if not kills the plants. Effective management for this pest is to use fipronil instead of malathion because it is more toxic and effective towards *T. varians* larvae even in low concentration (Basari et al., 2019). Meanwhile, their natural enemies were parasitoid ichneumonid wasps, *Enicospllus* sp. and *Goryphus* sp. that parasitized larvae and pupae of *T.varians* (Kedar et al., 2014).

Mites
Mites are small arachnids from the subclass Acari with four pairs of legs. Some dwelling mites on fig plants are beneficial, others cause no visible injury to plants, but some are serious plant pests (**Figure 5**). Other than insects, mite infestation also can cause a similar effect like sap-sucking insects to the plant vigor. However, sooty mold would not be present but attacked leaves by mites usually turn bronze or rusty with the presence of the web.

Fig blister mite *Aceria ficus* (Eriophyidae) from Suborder Prostigmata is a worldwide pest of fig which is present in Africa, Asia and North America. This mite colonized fig trees especially the buds and young leaves and is a vector to Fig Mosaic Virus. In addition to vectoring viruses, their feeding also can cause various types of discoloration (chlorosis) and even malformation as they can also stunt plants if an infestation is severe.

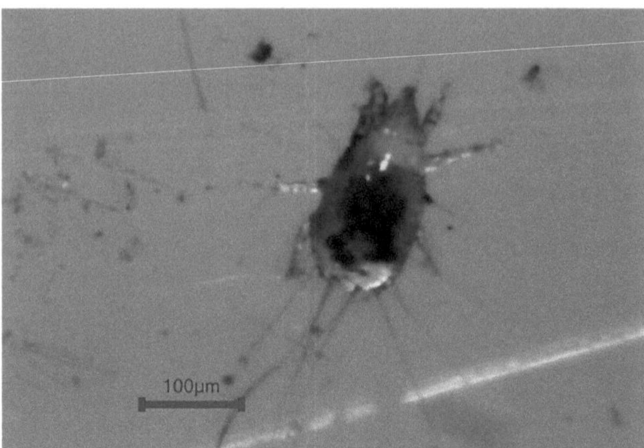

Fig. 5: Mold mite found on the remains of scale insects on fig leaf.

a) *Aceria ficus* Cotte

Ficus mottle mite *A. ficus* is a mite from the family Eriophyidae which is a plant parasite that has caused great economic loss to fig cultivation by producing symptoms like leaf mosaic, deformation of leaf, chlorosis, scarring and russeting of the eye scales, stunting, immature leaf drop, fruit mosaic, and fruit drop. The pest shows a high preference or hosts specific to plant species of the genus *Ficus*.

Their biological development is all year round with many life stages that can be found together. Adult mite measure is between 200 to 250 µm in length and have a very short life cycle of about 14 days comprising egg and two nymphal stages (Ashihara et al., 2004). This mite can multiply to high densities within a few weeks. Their dispersal is passive which is assisted mainly by wind currents. Many researchers have reported that Phytoseiidae mites such as *Amblyseius* spp., *Neoseiulus* spp. and *Phytoseiulus* spp., and Tydeidae mite, *Pronematus* sp., as natural enemies against *A. ficus*. While for chemical control, thiophanate-methyl was reported able to minimize *A. ficus* damage in the field (Ashihara et al., 2004).

DISEASES IN *Ficus* sp.

Several diseases have been reported to infect *Ficus* sp. which directly affect the development of the plant such as anthracnose, *Botrytis* blight, gray mold, rust, leaf spot and fruit rot.

Anthracnose

Anthracnose is one of the diseases that can attack fig plants, especially the leaf of *F. elastica* and fruit of *F. carica* (Choi et al., 2013). This disease is mainly caused by *Colletotrichum* sp. Figs that are being infected with this disease can develop early symptoms of small, slightly sunken and circular spots on the leaf or fruit surface. Over time, the lesions will become bigger and turn into necrotic spots resulting in semi-soft decay, premature release of fruit, or defoliation of leaves. Under humid conditions, the darkened lesion area can form a brown to the salmon-colored mass of spores. These spores can be circular to ellipsoid in shape with the presence of septae.

***Botrytis* blight**

Botrytis blight pathogen can infect the figs during the cool and damp season especially on the cutting parts. This disease is mainly caused by *Botrytis cinerea* and the infection normally starts on the young leaves and young stems. The infected leaves and stems will turn to brown or black color with the presence of gray mycelium on the surface. Eventually, the disease will result in the collapse of foliage.

Gray mold
Aside from *Botrytis* blight, *B. cinerea* can also infect the fig fruit during preharvest and postharvest. The symptoms of the infection are the development of brown-soaked fruit rot and fruit decay. Over time, the fruit will be covered with gray to brown-colored mycelium mass. The occurrence of this disease not only can degrade the quality of the fruit but also can affect the yield production. The presence of *B. cinerea* in fig can be distinguished by the formation of the ellipsoidal and oval shape of conidia and the presence of grayish to brown colored mycelium (Cheong et al., 2013).

Fig Rust
Ficus sp. often show rust symptoms (Verga & Nelson, 2014) compared to other host genera such as Maclura and Morus. It has been reported previously that fig rust is caused by the fungus *Cerotelium fici*. Its spores are produced in the pustule (Palmateer et al., 1999) on the lower leaves part (**Figure 6**). The characteristic of the spore indicates brown, single-celled and spiky ornamentation (Palmateer et al., 1999) on the cell surface. Basically, the rust spores develop well especially during warm and humid weather. The first symptom that appears on the leaves is a small and yellow spot (Palmateer et al., 1999) on leaves. Then, a large brownish spot formed with a reddish border. Soon after, reddish-brown pustules (Verga & Nelson, 2014) developed at the lower part of the leaves (Verga & Nelson, 2014). If the disease gets severe, the angular brown spot on the upper leaves forms which later forms a necrotic region. This leads to the defoliation of the tree and produces unripe fruit. In addition, the spores have the ability to survive even on the fallen diseased plant (Palmateer et al., 1999).

Fig. 6: Appearance of tiny and orange pustules on the abaxial side of the leaves.

***Cercospora* Leaf Spot**
Leaf spot is another type of disease that affects the leaves. It has been reported that the disease spreads around the continental United States (Schubert et al., 1999). The causal agent of leaf spot is *Cercospora*

fici. Basically, on the lesion surface, *Cercospora fici* forming threadlike conidia from conidial bearing hyphae infect the leaves (Palmateer et al., 1999). The first symptom indicates reddish-brown, angular to the irregular shape. When the spot enlarges, the center becomes tan with a dark brown margin surrounding the spot. During the summer rainy season, the most conducive environment, the disease developed well and the fungus grew rapidly (Schubert et al., 1999) which infects the leaves directly. Similar to rust disease, leaf spot leads to defoliation of leaf.

Fruit Rot
Mainly, this disease affects only fruit as no foliar symptom was shown and observed from the previous study in Montenegro (Latinović et al., 2014). The fungi to this disease are identified as *Alternaria alternate* (Latinović et al., 2014). This colony comprises brown septate hyphae, branched conidiophores and ellipsoid to ovoid shape. Initially, the symptom appears which is a small circular, necrotic and sunken spot near the ostiolar canal. The spots are later enlarged gradually (Latinović et al., 2014) which infected the whole fruit root.

CONCLUSION
Some cultural practices such as the use of insecticides, fungicides, obtaining plant materials from a clean source and examining plants for infestation before movement within or across regions have been introduced widely to prevent and reduce pest infestation and disease severity. Broad-spectrum insecticides and fungicides are frequently the only options in nurseries, considering the very low level of injury that is acceptable in domestic and international markets. However, high dependency on chemical control is detrimental to the natural *Ficus* macrofauna and microfauna. Therefore, understanding the pest and pathogen biology, economic impact, and sustainable management application helps in reducing the pest infestation and disease occurrence, thus preventing the outbreak to the new area.

REFERENCES

Alford, D.V. (2014). Pest of Fruit Crops: A Colour Handbook. 2nd Ed. CRC Press, Taylor and Francis Group, Florida USA. 462 pp.

Aksit, T., Cakmak, I. & Moser, J. (2007). Attack by *Pyemotes johnmoseri* (Acari: Pyemotidae) on *Hypoborus ficus* (Coleoptera: Scolytidae) in fig trees in Turkey. Experimental and Applied Acarology, 41: 251-254.

Aksit, T., Cakmak, I. & Ozsemerci, F. (2005). Some new xylophagous species on fig trees (*Ficus carica* cv. Calymirna L.) in Aydin, Turkey. Turkish Journal of Zoology, 29: 211-215.

Arthurs, S., Chen, J., Dogramaci, M, Ali, A.D. & Mannion, C. (2011). Evaluation of *Montandoniola confusa Streito* and *Matocq* sp. nov. and *Orius insidiosus* Say (Heteroptera: Anthocoridae), for control of *Gynaikothrips uzeli* Zimmermann (Thysanoptera: Phlaeothripidae) on *Ficus benjamina*. Biological Control, 57: 202-207.

Arya, P.V. (2020). Recent diversity and potential biological control studies on major ornamental *Ficus* sp. defoliating moth bombycid *Trilocha* (=*ocinara*) *varians* (Walker) (Lepidoptera: Bombycidae). Journal of Experimental Zoology India,23(1): 215-217.

Ashihara, W., Kondo, A., Shibao, M., Tanaka, H., Hiehata, K. & Izumi, K., (2004). Ecology and control of Eriophyid mites injurious to fruit trees in Japan. Japan Agricultural Research Quarterly (JARQ), 38 (1): 31-41.

Basari, N., Mustafa, N.S., Yusrihan, N.E.N., Chin, W.Y. & Ibrahim, Z. (2019). The effect of temperature on the development of *Trilocha varians* (Lepidoptera: Bombycidae) and control of the Ficus plant pest. Tropical Life Sciences Research, 30(1): 23-31.

Basset, Y. & Novotny, V. (1999). Species richness of insect herbivore communities on *Ficus* in Papua New Guinea. Biological Journal of the Linnean Society, 67: 477-499.

Berning, D., Dawson, G. & Foggia, M. (2014). "*Planococcus ficus*" (On-line), Animal Diversity Web. Accessed March 30, 2021 at https://animaldiversity.org/accounts/Planococcus_ficus

Bragard, C., Dehnen-Schmutz, K., Di Serio, F., Jacques, M., Miret, J.A.J., ...& Gonthier, P. (2021). Commodity risk assessment of *Ficus carica* plants from Israel. EFSA Journal, 19(1): 6353, DOI: 10.2903/j.efsa.2021.6353

Cheong, S.S., Choi, I.Y. & Lee, W.H. (2013). Occurrence of gray mold caused by *Botrytis cinerea* on common fig in Korea. The Korean Journal of Mycology,41(1): 38-41.

Choi, I.Y., Park, J.H., & Shin, H.D. (2013). First confirmed report of anthracnose fruit rot caused by *Colletotrichum gloeosporioides* on common fig in Korea. Plant Disease, 97(8): 1119-1119

David, B.V. & Regu, K. (1995). *Aleurodicus dispersus* Russell (Aleyrodidae: Homoptera) a whitefly pest new to India. Pestology XIX (3): 5-7.

Dogramaci, M., Arthurs, S.P., Chen, J., McKenzie, C., Irrizary, F., & Osborne, L. (2011). Management of chilli thrips *Scirtothrips dorsalis* (Thysanoptera: Thripidae) on peppers by *Amblyseius swirskii* (Acari: Phytoseiidae) and *Orius insidiosus* (Hemiptera: Anthocoridae). Biological Control, 59: 340-347.

Flaishman, M.A., Rodov, V., & Stover, E. (2008). The Fig: Botany, Horticulture, and Breeding. Horticultural Reviews, 34: 113-196.

Funderburk, J.E., Denmark, H.A., Mound, L.A., Skarlinsky, T.L. & Mannion C. (2014). Leaf-gall thrips of Ficus, *Gynaikothrips ficorum* (Marchal) and *Gynaikothrips uzeli* (Zimmerman). University of Florida Featured Creatures Publication Number EENY-324.

Gencer, N.S., Coskuncu, K.S. & Kumral N.A. (2007). The colonization preference and population trends of larval fig psylla, *Homotoma ficus* L. (Hemiptera: Homotomidae). Journal of Pest Science, 80: 1-8.

Hodel, D.R. (2017). New Pest of Landscape Ficus in California. CAPCA ADVISORS: 58-62.

Hodel, D.R., Arakelian, G., Ohara, L.M., Wilen C. & Dara, S.K. (2016). The ficus leaf-rolling psyllid: a new pest of *Ficus microcarpa*. PalmArbor, 2016-2: 1-9.

Kamarubahrin, A.F., Haris, A., Abdul Shukor, S., Mohd Daud, S.N., Ahmad, N., Zulkefli, Z., Muhamed, N.A. & Abdul Qadir, A.H.M. (2019). An overview Malaysia as a hub of planting prophetic fruits. Malaysian Journal of Sustainable Agriculture, 3(1): 13-19.

Kedar, S.C., Kumaranag, K.M. & Saini, R.K. (2014). First report of *Trilocha* (=*ocinara*) *varians* and its natural enemies on *Ficus* spp. from Haryana, India. Journal of Entomology and Zoology Studies, 2(4): 268-270.

Khaustov, A.A. (2004). New replacement names for *Pyemotes moseri* Khaustov, 1998 and *Archidispus hrosui* Khaustov, 2004 (Acari: Heterostigmata: Pyemotidae, Scutacaridae). Acarina, 12: 47.

Kumar, V., Seal, D.R. & Kakkar, G. (2009). Chilli thrips *Scirtothrips dorsalis* Hood (Insecta: Thysanoptera: Thripidae). IFAS Extension University of Florida. https://edis.ifas.ufl.edu/in833

Latinović, N., Radišek, S., & Latinović, J. (2014). First report of *Alternaria alternata* causing fruit rot on Fig (*Ficus carica*) in Montenegro. Plant Disease, 98(3): 424-424.

Mawa, S., Husain, K., & Jantan, I. (2013). *Ficus carica* L. (Moraceae): Phytochemistry, Traditional uses and biological activities. Evidence - Based Complementary and Alternative Medicine, 1-8. https://doi.org/10.1155/2013/974256

Moniruzzaman, M., Yaakob, Z., Khatun, R. & Awang, N. (2017). Mealybug (Pseudococcidae) infestation and organic control in fig (*Ficus carica*) orchards of Malaysia. Biology and Environment: Proceedings of the Royal Irish Academy 2017. https://doi.org/10.3318/BIOE.2017.04

Nair, S., Senior, R., Umeda, K., Ellsworth, P., Schuch, U., Gouge, D. & Li, S. (2015). Integrated Pest Management of the Ficus Whitefly. Cooperative Extension Arizona Pest Management Centre, University of Arizona. https://cals.arizona.edu/apmc/docs/Ficus_Whitefly_Short2.pdf

Navasero, M. M. & Navasero, M. V. (2014). Biology of *Trilocha varians* (Walker) (Lepidoptera: Bombycidae) on *Ficus benjamina* L. inthe Philippines. Philippine Entomologist28(1): 43-56.

Palmateer, A. J., Tarnowski, T., & Roberts, P. D. (1999). Florida Plant Disease Management Guide: Fig (*Ficus carica*). University of Florida, Institute of Food and Agricultural Sciences. 3 p.

Riley, D.G., Joseph, S.V., Srinivasan, R. & Diffle, S. (2011). Thrips vectors of tospoviruses. Journal of Integrated Pest Management, 1(2). https://doi.org/10.1603/IPM10020

Schubert, T. S., El-Gholl, N. E., & Alfieri, Jr, S. A. (1999). Cercospora Leaf Spot of Fig. Plant Pathology Circular No. 394, Florida Department of Agriculture and Consumer Services Division of Plant Industry.

Scudder, G.G.E. & Canning R.A. (2005). The Coleoptera families of British Columbia. The Ministry of Forests, Lands, Natural Resource Operations and Rural Development. British Columbia. 110 pp. https://www.for.gov.bc.ca/hfd/library/FIA/2005/FIA2005MR118b.pdf

Verga, A., & Nelson, S. (2014). Fig Rust in Hawai'i. Plant Disease. College of Tropical Agriculture and Human Resources, University of Hawai'i at Mānoa. Fromhttps://www.ctahr.hawaii.edu/oc/freepubs/pdf/PD-100.pdf

Zada, A., Dunkelblum, E., Assael, F., Harel, M., Cojocaru, M. & Mendel, Z. (2003). Sex pheromone of the vine mealybug, *Planococcus ficus* in Israel: occurrence of a second component in a mass-reared population. Journal of Chemical Ecology, 29: 977-988.

Techniques for Extraction of Bioactive Compounds from Fig (*Ficus* spp.): A Review

Sahena Ferdosh*, Nadirah Abd Rahim
¹Department of Plant Science, Kulliyyah of Science, International Islamic University Malaysia (IIUM), 25200 Kuantan, Pahang, Malaysia
Corresponding author email: sahena@iium.edu.my

ABSTRACT

Fig (*Ficus* spp.), an important Asian crop is now widely grown throughout the world for its nutritional and therapeutic properties. This paper explores different methods of phytochemical screening and extraction techniques of phytochemicals constituents from common species of *Ficus* obtained from all accessible library databases and electronic search engine. The extraction of phytochemicals from various *Ficus* species by different conventional and non-conventional techniques have been reported. Soxhlet, maceration and microwave-assisted extraction are among the methods reported to isolate compounds from various species of *Ficus*. Common phytochemical groups found in the *Ficus* research are alkaloids, flavonoids, coumarins, phenolics, saponin and terpenoids which distinctively based on the plant parts, species, and the solvent used in the study. However, non-conventional green extraction technologies on *Ficus* study are still limited. Thus, this review might provide a clearer future perspective to the readers.

Keywords: Conventional and non-conventional, Extraction techniques. *Ficus* spp, Phytochemicals screening.

INTRODUCTION

Medicinal plants have long been used to cure a variety of diseases all around the globe. World Health Organization (WHO) stated that eighty percent of the people in developing countries use traditional medicines as their primary treatment or prevention of diseases (Velavan, 2015). Nowadays, natural products from plants do not only meet the primary healthcare needs of most of the population in developing countries, but they are also gaining interest in developed countries due to rising health-care costs. *Ficus* species is not left behind since it is also among the famous medicinal plants. *Ficus* which belongs to Moraceae family, is one of the largest group of angiosperms that comprises more than 800 species of scrubs, trees, hemi epiphytes, climbers, and creepers and widely distributed in the tropics and subtropics throughout the world (Mawa et al., 2013).

Commonly studied *Ficus* species reported by Sirisha et al. (2010) are *Ficus benghalensis* (Indian banyan), *Ficus carica* (commonly known as edible fruit), *Ficus sycomorus, Ficus thonningii, and Ficus regiliosa* (sacred fig). Different parts of these species such as the leaves, roots, stem barks, fruits, and latex are medicinally used in different health treatments and diseases. For instances, the leaves of *F. carica* were reported to be effective as anti-inflammatory agents for conditions such as painful or swollen piles, insect sting and bites (Ali et al., 2012). Other parts of the *F. carica* and *F. sycomorus* such as the tree bark, the latex, the different stages of ripening fruits and the young shoots have been used to treat tumors and diseases associated with inflammation (Lansky et al., 2008).

All those potential biological properties are derived from the phytochemicals present in the plant. Many studies have been conducted to identify the components in *Ficus* since many decades ago. The present of important phytochemicals such as flavonoids, phenolics, terpenoids, and alkaloids in *Ficus* are detected by various phytochemicals screening tests, accordingly to the different groups of phytochemical compounds (Velavan, 2015). Moreover, in order to isolate the phytochemicals from the *Ficus* spp, different methods of extraction have been conducted by the researchers previously (Zhang et al., 2018). The common conventional method of extractions used by most researchers to extract bioactive compounds from plant are maceration, decoction, and Soxhlet extraction. However, recently, most researchers have shifted to the green technologies such as microwave-assisted extraction (MAE), ultrasonic-assisted extraction (UAE), and supercritical fluid extraction (SFE) methods which gave more advantages in terms of extraction's duration, extract quality and environmental safety (Mohammed & Al-Saddi, 2003; Azwanida, 2015; Zhang et al., 2018; Sharma et al., 2020). However, due to the vast species of *Ficus*, the studies of these phytochemicals and extraction methods on specific species are still lacking. Therefore, this review aimed to give bright perspective for future on the phytochemical screening, phytochemical constituents and the extraction method on common species of *Ficus*.

PHYTOCHEMICAL SCREENING OF *Ficus* spp.

Phytochemicals such as carbohydrates, lipids, terpenoids, phenolics and alkaloids are constituents that produced by plants and are believed to bring benefits to human health due to the biological activities of the specific constituents. Phytochemicals found in plants have been reported to serve many functions in plant itself such as pathogen defense, plant protection, plant growth and plant signaling (Huang et al., 2016). Meanwhile, bioactive phytochemicals are defined as those that able to interact with other component of living tissue which led to various of possible biological effects (Guaadaoui et al., 2014; Huang et al., 2016). In order to identify the phytochemicals in plant, researchers always started with the preliminary screening test and then move on to the analysis of secondary metabolites using more advanced technologies (Khoo et al., 2018). In past decades, the presence or absence of phytochemicals in various parts of *Ficus* plants have been determined by various methods specifically to the distinct groups of phytochemicals. According to Sirisha et al. (2010), flavonoids, alkaloids, terpenoids, phenolics, steroids, saponins and coumarins were the constituents that commonly found in different *Ficus* species.

Screening of Alkaloids

Alkaloids are a large group of organic compounds by nature which contain nitrogen atom in their arrangement that contributes alkalinity to these compounds. In terms of their structures, alkaloids are divided into different classes like indoles, quinolines, isoquinolines, pyrrolidines, pyridines, pyrrolizidines, tropanes, terpenoids and steroids (Farnsworth, 1966). Plant's age, habitat, part tested, season, time of harvest, chemical races of plants and susceptibility of alkaloid form to reagents, may all influence the results of alkaloid testing of plant material (Panche et al., 2016). Several methods for alkaloids screening are Bouchardat, Dragendorff, Ecolle, Gold Chloride, Hager, Kraut, Marme, Mayer, Platinum chloride, Scheibler, Sonnenschein, Valscr, and Wagner's test. In previous study by Sharma & Sajgotra (2017), screening test such as Dragendorff's test, Mayer's test, Wagner's test, and Hager's test are used to determine the presence of alkaloids in the fruit's extract of *F. carica* as stated in Table 1. In addition, Coker & Oaikhena (2020) and Uku et al. (2020) reported that alkaloids are also present in *F. thonningii* extract (Table 2). In other study, different extracts of stem bark of *F. sycomorus* are tested for the present of alkaloids and resulted that those alkaloids were found in aqueous, chloroform, ethyl acetate, methanol extracts but absent in the hexane extract of stem bark of *F. sycomorus* (Oluwasesan et al., 2013) (Table 3).

Screening of Flavonoids

Flavonoids are an essential group of bioactive compounds that contains polyphenolic structure that can be divided into subgroups such as isoflavones, neoflavonoids, flavones, flavonols, flavanones, flavanonols, flavanols or catechins, anthocyanins and chalcones depending on the structural features of the C ring (Panche et al., 2016). Many flavonoids have evolved as secondary metabolites with the properties like antimicrobial, insecticidal, and pharmacological due to the interaction of compounds with nucleic acid or proteins thus increased the researcher's interest because of the therapeutic effects (Mawa et al., 2013). In a study by Sharma & Sajgotra (2017), Shinoda's test was used to determine the presence of flavonoids in the fruit's extract of *F. carica*. However, in their study, flavonoids are found only in the ethanol extract of *F. carica's* fruit. On the other hand, all five *F. sycomorus* different plant's stem bark extracts namely aqueous, chloroform, ethyl acetate, hexane and methanol extracts are revealed to contain flavonoids (Table 3) (Oluwasesan et al., 2013).

Screening of Glycosides and other Groups of Phytochemicals
An abundance of glycosides and other phytochemicals screening studies on many species of *Ficus* has been reported previously. Generally, Bontrager's test, Keller-Killiani test, Baljet's test and Legal test are the phytochemical screening methods that have been used to test the presence of glycosides meanwhile Benedict's test, Molisch's test and Fehling's test were used to identify the present of carbohydrates and sugar (Sharma & Sajgotra., 2017). They reported that several groups of phytochemicals such as glycosides, phenolics, tannins, fats and carbohydrates are present in *F. carica*'s leaves extract. In this investigation, fats and oils are the only constituents found in the petroleum ether extracts of the leaves of *F. carica* whereas, tannins, and polyphenol are found to be present in chloroform extracts of the leaves (Sharma & Sajgotra, 2017). Moreover, most phytochemicals found in ethanolic extract of the leaves of *F. carica* are carbohydrates, glycosides, flavonoids, tannins, and polyphenol as stated in Table 1.

The aqueous extract of *F. thonningii* fruit consists of polyphenols, flavonoids, saponins, tannins, and betacyanin as mentioned in Table 2. In addition, Coker & Oaikhena (2020) reported that terpenoids, anthraquinones, saponins, tannins and are present in *F. thonningii* crude extract, while cardiac glycosides and steroids are found to be absent. More phytochemicals such as tannins, resins and cardiac glycosides reported by Oluwasesan et al. (2013) are displayed in Table 3. These preliminary phytochemicals studies are important as they help in detecting the bioactive properties found in medicinal plants, which may contribute to drug discovery and medicinal product development.

Table 1: List of major phytochemical groups in different extracts of *F. carica* and screening tests used for detection.

Phytochemical	Aqueous extract of ripe dried fruit (Mawa et al., 2013)	Petroleum ether leaves extract (Sharma & Sajgotra., 2017)	Chloroform leaves extract (Sharma & Sajgotra., 2017)	Ethanol leaves extract (Sharma & Sajgotra., 2017)	Hydroalcohol (50% ethanol + 50% H_2O) leaves extract (Ali et al., 2012)
Alkaloids	+ (Hager's test)	- (Hager's & Mayer's test)	+ (Hager's & Mayer's test)	+ (Baljet's test)	-
Flavonoids	+	-	-	+ (Shinoda's test)	+
Carbohydrates	-	-	-	+ (Benedict's test)	+
Tannins	-	-	+ (Lead acetate test)	+ (Lead acetate test)	+
Glycosides	-	-	-	+ (Molisch's, Benedict's & Borntragers test)	-
Steroids	-	-	-	-	+

Present (+), Absent (-)

Table 2: List of major phytochemical groups in different extracts of *F. thonningii* and screening tests used for detection

Phytochemicals	Aqueous extract of fruits (Uku et al., 2020)		Crude extract of *F. thonningii* (Coker & Oaikhena, 2020)	
Alkaloids	-	(Mayer's test)	+	(Dragendorff's test)
	-	(Hager's test)		
Flavonoids	+	(Shinoda test)	-	
	+	(NaOH test)		
Total polyphenols	+	(Lead acetate)	-	
Tannins	+	(Cu citrate)	+	
Saponins	+	(FeCl3)	+	
Steroids	-	(Liebermann-Burchard)	-	
Resins	-	(Acetic acid + H_2SO_4)	-	
Betacyanins	+	(NaOH)	+	
Anthraquinones	x		+	
Cardiac glycosides	x		-	(Ferric chloride)

Present (+), Absent (-), Not tested (x)

Table 3: Major phytochemical group found in different stem bark extracts of *F. sycomorus* and screening tests used for detection (adapted from Oluwasesan et al., 2013)

Phytochemicals	Aqueous stem bark extract	Chloroform stem bark extract	Ethyl acetate stem bark extract	Methanol stem bark extract	Hexane stem bark extract
Alkaloids	+ (Wagner's & Hager's test)	+ (Wagner's & Hager's test)	+ (Wagner's & Hager's test)	+ (Wagner's & Hager's test)	-
Flavonoids	+ (NH4OH test)	+ (NH4OH test)	+ (NH4OH test)	+ (NH4OH test)	+ (NH4OH test)
Reducing Sugar	+ (Fehling's test)	+ (Fehling's test)	-	+ (Fehling's test)	-
Tannins	+	-	+	+	+
Cardiac glycosides	-	+	-	-	+
Saponins	+	+	+	+	+
Steroids	-	-	-	-	-
Resins	-	-	+	-	-

Present (+), Absent (-)

BIOACTIVE PHYTOCHEMICAL CONSTITUENTS OF *Ficus* spp.

Wide variety of bioactive compounds found in *Ficus* species and wide biological properties have made them among the most popular medicinal plant. Starting from the preliminary phytochemical screening followed by the identification of specific compounds using various technologies have been done to determine the presence of chemicals compound that might possess therapeutic effects on living cells. In previous studies, various bioactive compounds have been found in different species of *Ficus*. Table 4 shows the bioactive compounds found few species of *Ficus* which are *F. carica, F. thonningii, F. lacor* and *F. exasperata.*

Table 4: Different bioactive compounds reported in *Ficus* spp.

Species	Bioactive Compounds	References
F. carica	Gallic acid dipentoside, dihydroxybenzoic acid hexoside pentoside, dihydroxybenzoic acid pentoside, dihydroxybenzoic acid dipentoside, syringic acid hexoside, syringic acid malate, dihydroxybenzoic acid hexoside, vanillic acid glucoside, dihydrobenzoic acid, hydrobenzoic acid, vanillic acid, caffeoylmalic acid, chlorogenic acid, 5-O-caffeoylquinic acid psoralic acid glucoside, ferulic acid malate, coumaroyl hexoside, sinapic acid, acid hexoside, coumaroylmalic acid, sinapic acid malate, rutin, isoquercetin, astragalin, taxifolin, galangin, myricetin, kaempferol, quercetin 3-o-malonylglucoside, luteolin, vitexin, isoschaftoside, 7,4-dihydroxyflavone, 7-hydroxyflavone, 7,4,3-trihydroxyflavone, 5,3,4-trihydroxyflavone, 5,7-dihydroxyflavone, isoorientin, orientin, cynaroside, apigenin, 4-hydroxyflavanone, 2-hydroxyflavanone, 7-hydroxyflavanone, hesperetin, naringenin, catechin, prenylhydroxygenistein, prenylgenistein, formononetin, hydroxy-dimethoxyisoflavone, daidzein, genistein, cajanin, biochanin A, murrayacarpin B, esculetin hexoside, dihydroxycoumarin, umbelliferone, phellodenol A, prenyl-7hydroxycoumarin, psoralic acid, marmesinin, 4′,5′dihydropsoralen, hydroxypsoralen, isopentenoxypsoralen, hydroxypsoralen hexoside, oxypeucedanin hydrate, psoralen, methoxypsoralen, bergapten, oxypeucedanin, prenyl methoxypsoralen, limonene, menthol, α-guajene, α-cubebene, α-ylangene, copaene, β-bourbonene, β-elemene, α-gurjunene, β-caryophyllene, β-cubebene, alloaromadendrene, α-caryophyllene, δ-muurolene, D-germacrene, ledene, δ-elemene, δ-cadinene, α-muurolene, 3-methyl-butanal, 2-methyl-butanal, (E)-2-pentenal, hexanal, (E)-2-hexenal, 1-penten-3-ol, 3-methyl-1-butanol, 2-methyl-1-butanol, 1-heptanol, benzyl alcohol, (E)-2-nonen-1-ol, phenylethyl alcohol, 3-pentanone, β-cyclocitral, methyl butanoate, methyl hexanoate, hexyl acetate, ethyl benzoate,	Li et al., 2021; Mawa et al., 2013

	methyl salicylate, bauerenol, calotropenyl acetate, lupeol acetate, oleanolic acid, methyl maslinate, lupeol, 24-methylenecycloartanol, β-sitosterol, palmitic acid methyl ester, palmitic acid ethyl ester, palmitic acid, linolenic acid methyl ester, linoleic acid, linoleic acid ethyl ester psoralen, stearic acid ethyl ester.	
F. thonningii	Thonningiol, thonningiisoflavone, β-sitosterol, β-sitosterol 3-O-b-D-glucopyranoside, gancaonin, bamyrin acetate, friedelin, lupeol hexanoate, lupeol acetate, alpinumisoflavone, wighteone, dehydroferreirine, β-isoluteone, taxifolin, lupiwitheone hydrate, p-menthane-3,6-diol, conrauiflavonol, aromadendrin, shuterin, luteone, hydroxyalpinumisoflavone	Fongang et al., 2015
F. lacor	B-sitosterol, lupeol, stigmasterol, campesterol, α-amyrin, β-amyrin, scutellarein glucoside, scutellarein, infectorein, sorbifolin, bergaptol, bergapten, α-hydroxyyursolic acid, oleanolic acid, protocatechuic acid, ursolic acid, maslinic acid, alanine, tyrosine, methionine, lactose, sucrose, galactose, vitamin C, β-carotene	Ghimire et al., 2020
F. exasperata	α-terpinene, α-pinene, β-pinene, α-thujene, sabinene, 2-carene, α-phellandrene, limonene, (Z)-β-ocimene, β-myrcene, (E)-β-ocimene, linalool, verbenone, β-cyclocitral, bornyl acetate, eucalyptol, α-cubebene	Oladosu et al., 2009

EXTRACTION METHODS OF PHYTOCHEMICALS FROM *Ficus* spp.

Extraction is the primary step to isolate desired compounds from plants. Theoretically, extraction is defined as the process of separating medicinally active plant's compounds by specific solvent with standardized procedure (Azwanida, 2015). Various methods of extraction have been used to isolate the bioactive compounds from *Ficus* ranged from conventional to non-conventional. Many factors affecting the extraction process such as the pre-extraction preparation of sample, solvent type, temperature, and duration of extraction (Mohammed & Al-Saddi, 2003). Moreover, the isolation of specific compounds might be varied depending on the extraction process. Table 5 displays several extractions method that have been used to extract bioactive compounds from *Ficus* spp.

Table 5: Effect of extraction methods and solvent on yield percentage of phytochemicals in *Ficus* spp.

Extraction method	Plant species (Part of plant)	Extraction Solvent	Yield percentage (%)	References
Maceration	*Ficus religiosa* (leaves)	Ethanol	26.52	Charde et al. (2010)
Maceration	*Ficus thonningii* (fruits)	H_2O	8.19	Uku et al. (2020)
Maceration	*Ficus carica* (latex) *Ficus sycomorus* (latex)	Methanol	NS	Abdel-Aty et al. (2019)
Soxhlet	*Ficus sycomorus* (leaves)	n-hexane	5.4	Mudi et al. (2015)
		Chloroform	2.2	
		Ethyl acetate	1.3	
		Methanol	3.9	
Soxhlet	*Ficus sycomorus* (fruits	N-hexane	3.5	Mudi et al. (2015)
		Chloroform	1.3	
		Ethyl acetate	1.6	
		Methanol	1.2	
Soxhlet	*Ficus carica* (leaves)	Methanol	NS	Eteraf-Oskouei et al. (2015)
Soxhlet	*Ficus glomerata*	Petroleum ether	1.95	Supriya et al. (2012)
		Methanol	8.96	
Microwave-assisted extraction	*Ficus racemosa* (fruits)	Acidified solvent	NS	Sharma et al. (2020)
Microwave-assisted extraction	*Ficus carica* (fruits)	Acidified solvent	NS	Backes et al. (2018)
Microwave-assisted hydrodistillatio	*Ficus deltoidea* (leaves)	Water	NS	Mohd-Nasir et al. (2020)
Heat-assisted extraction	*Ficus carica* (fruits)	Acidified solvent	NS	Backes et al. (2018)
Ultrasonic-assisted extraction	*Ficus carica* (fruits)	Acidified solvent	NS	Backes et al. (2018)

*NS=Not specified.

Conventional Extraction Methods

Maceration

Maceration is among the oldest extraction methods to extract phytochemicals from *Ficus* spp. (Azwanida, 2015; Ammar et al., 2017; Abdel-Aty et al., 2019; Uku et al., 2020), because of maceration is the simplest method with least cost of instrumentations (Zhang et al., 2018). However, maceration process takes a long extraction time and low efficiency of extraction (Azwanida., 2015; Zhang et al., 2018). Moreover, large volume of solvent is used per extraction process, which needed a proper management to dispose the waste (Azwanida, 2015). In a previous study by Charde et al. (2010), the leaves of *Ficus religiosa* were macerated using ethanol and resulted in 26.52% of yield as stated in Table 5. Meanwhile, Uku et al. (2020) reported that the yield percentage of water maceration of *F. thonningii* fruits was 8.19 %. These also showed that different solvent extraction used resulted to a different yield percentage of extraction. The extraction of targeted compounds using maceration method is very dependent on the extraction solvent used (Mohd-Nasir et al., 2020).

Soxhlet

Conventional Soxhlet extraction is a method that apply heat to the solvent in order to extract the bioactive compounds from the sample. In comparison to maceration method, this method needs less solvent. However, Soxhlet extraction has drawbacks, including the risk to dangerous and flammable solvent, as well as the tendency of toxic emissions during the process (Azwanida, 2015). Mudi et al. (2015) reported the extraction of phytochemicals in leaves of *F. sycomorus* using distinct solvents to identify the efficiency of each solvent used. Among the four solvents used, the highest yield recorded was 5.4% obtained from n-hexane extraction, meanwhile the least yield was 1.3% obtained from ethyl acetate Soxhlet extraction. On another study using Soxhlet extraction method by Eteraf-Oskouei et al. (2015) revealed that the methanolic extract of *F. carica* leaves showed anti-inflammatory activities at the levels of cell migration, exudates, tissue weight, and content of proinflammatory mediators. Different studies on *Ficus* using Soxhlet method are listed in the Table 5.

Non-conventional Extraction Methods

Microwave-assisted Extraction

Microwave-assisted extraction (MAE) has recently attracted many of researchers due to its benefits, such as the use of less solvent and a shorter extraction time than other conventional extraction methods (Sharma et al., 2020). MAE has been shown to be more effective than traditional methods like Soxhlet extraction and solid-to-liquid extraction in several studies. However, the efficiency of MAE depends on the solvent's pH, microwave power, and extraction time. Thus, an optimization of MAE process for the maximum recovery of phytochemicals in the fruits of *F. racemosa* was studied by Sharma et al. (2020). The optimum conditions found in this study were 3.5 solvent pH, 360.55 W microwave power and 30.01 s of extraction time with 0.81 desirability. The extract was further analyzed, and it was found that ascorbic acid, gallic acid, catechin, tannic acid, chlorogenic acid, ferulic acid and quercetin were present with the amount 31.19 ± 1.05 mg/100 mL; 23.49 ± 0.91 mg/100 mL; 14.06 ± 1.19 mg/100 mL; 50.86 ± 1.3 mg/100 mL; 22.18 ± 1.08 mg/100 mL; 36.96 ± 1.4 mg/100 mL, respectively.

Ultrasonic-assisted Extraction

Ultrasonic-assisted extraction (UAE), or also known as sonication extraction, is an extraction method that uses ultrasound ranging from 20 kHz to 2000 kHz (Azwanida, 2015). In this method, the physical and chemical properties of sample are changed by disrupting the plant's cell wall causing the release of phytochemicals and allow the mass flow of solvent in the plant cells (Azwanida., 2015). A study of extraction efficiency of polyphenols between UAE and maceration by Jovanović et al. (2017) showed that UAE gave a higher amount of flavonoids yield based on the polyphenols content. In a study by Backes et al. (2018), the extraction of anthocyanin from the fruits of *F. carica* was done by using MAE, UAE and heat-assisted extraction method. It was found that UAE was the most effective method which produced 3.8 mg C/g P dw and 0.95 g R/g P dw, with an anthocyanin content of 19.4 mg C/g R obtained at conditions of 21.34 min, 310.58 W, ethanol, and 183.01±22.82 g/L.

Supercritical Fluid Extraction
Supercritical fluid extraction (SFE) is a modern method of extraction that currently getting researchers' attention due to its advantages such as more environmentally safe and provides shorter extraction time compared to other traditional methods (Azwanida., 2015). The major advantage of this extraction method is a slight modification of its condition such as slight reduction in temperature and pressure can result in almost entire solute precipitating out (Sapkale et al., 2010). SFE extracts retain or surpass the bioactivity of extracts obtained using traditional methods, according to many studies. This is because SFE process is unique in that it facilitates selective extraction which results an extract rich in targeted compounds and free of organic solvents as it uses only carbon dioxide. SFE method also reduces the loss of compounds due to degradation or reaction (Pereira & Meireles, 2009). Since there is no specific study of *Ficus* spp on this method, this may allow the researchers to discover and look further into this method for *Ficus* spp extractions of the desired compounds.

EFFECT OF EXTRACTION METHODS ON BIOACTIVITIES OF PLANT EXTRACTS
Numerous biological activities have been reported from various species of *Ficus* since many decades ago. Antimicrobial, antibacterial, antifungal, anticancer, antidiabetic, antioxidant, and hepatoprotective are among the bioactivities revealed from *Ficus* spp (Abdou et al., 2021; Li et al., 2021). These biological activities are mainly due to the presence of bioactive compounds in the distinct species of *Ficus*. However, several studies have found that the extract yield as well as the bioactivities of plant extracts vary depending on extraction methods (Hayouni et al., 2007). The isolation of bioactive compounds responsible for the bioactivities of plants are greatly dependent on the extraction process including the conditions and solvent used. Thus, the extraction method is a crucial factor in all situations which can be so challenging due to the instability of the desired compounds.

Conventional extraction methods such as maceration, decoction, percolation and Soxhlet extraction are often used as models to compare the efficacy of non-conventional extraction methods such as MAE, SFE, and UAE. A study by Osorio-Tobón (2020) on the non-conventional extraction of phenolic compounds that exhibit antioxidant and antimicrobial properties resulted that MAE, with the highest Total Phenolic Content (TPC) and the shortest extraction times, followed by UAE and SFE. In some cases, the traditional approach outperforms modern methods in terms of extraction yield. However, the non-conventional extraction methods always had a greater number of secondary metabolites that may possess various of biological activities (Mohd-Nasir et al., 2020).

According to Li et al. (2012) traditional solvent-based methods had lower antioxidant activity and phenolic content than MAE. By measuring ferric reducing antioxidant power (FRAP), oxygen radical absorbance capacity (ORAC), and TPC, the researchers were able to confirm that MAE was more efficient at raising antioxidant activity. In another study by Mohd-Nasir et al. (2020) on *F. deltoidea* leaves, microwave-assisted hydrodistillation and a conventional method were used to extract vitexin, a compound that have wide range of bioactivities including antioxidant, anticancer, anti-inflammatory and anti-hyperalgesic. The study revealed that vitexin obtained from non-conventional method was higher with 33.40 ± 0.98% (w/w) compared to conventional method which was only 15.0% (w/w). However, MAE efficiency can be influenced by few factors including the temperature of extraction, the solvent, and the time of extraction (Altemimi et al., 2017). Overall, the non-conventional extraction method with best parameters will always give a better plant extract with more potential bioactivities over the old conventional extraction method.

CONCLUSION

This paper reviewed the phytochemical screening methods, bioactive compounds and extraction methods of common species of *Ficus*. Various phytochemical constituents have been isolated from the leaf, stem, bark, fruits, and including the latex of *Ficus*. The principal phytochemicals of *Ficus* are flavonoids, polyphenols, glycosides and some water-soluble vitamins. The presence of these phytochemicals makes *Ficus* a medicinal plant which possesses various biological activities. Highest crude extracts were observed in maceration followed by Soxhlet method. No complete study on the optimization of total extracts, identification of bioactive compounds and their therapeutic properties were reported for non-conventional method for *Ficus* sp. However, SFE which is regarded as green technology could be further used as a non-conventional technique to extract bioactive compounds and discover their therapeutic properties from *Ficus* spp.

REFERENCES

Abdel-Aty, A. M., Hamed, M. B., Salama, W. H., Ali, M. M., Fahmy, A. S., & Mohamed, S. A. (2019). *Ficus carica, Ficus sycomorus* and *Euphorbia tirucalli* latex extracts: Phytochemical screening, antioxidant and cytotoxic properties. *Biocatalysis and Agricultural Biotechnology, 20,* 101199.

Abdou, R., Mojally, M., & Attia, G. H. (2021). Investigation of bioactivities of endophytes of *Ficus carica* L. Fam Moraceae. *Bulletin of the National Research Centre, 45*(1). doi:10.1186/s42269-021-00505-1

Ali, B., Mujeeb, M., Aeri, V., Mir, S. R., Faiyazuddin, M., & Shakeel, F. (2012). Anti-inflammatory and antioxidant activity of *Ficus carica* Linn. leaves. *Natural Product Research, 26*(5), 460–465.

Altemimi, A., Lakhssassi, N., Baharlouei, A., Watson, D. G., & Lightfoot, D. A. (2017). Phytochemicals: Extraction, isolation, and identification of bioactive compounds from plant extracts. *Plants, 6*(4), 42.

Ammar, I., BenAmira, A., Khemakem, I., Attia, H., & Ennouri, M. (2017). Effect of Opuntia *Ficus-indica* flowers maceration on quality and on heat stability of olive oil. *Journal of Food Science and Technology, 54*(6), 1502–1510.

Ammar, I., Ennouri, M., & Attia, H. (2015). Phenolic content and antioxidant activity of cactus (*Opuntia ficus-indica* L.) flowers are modified according to the extraction method. *Industrial Crops and Products, 64, 97–104.*

Azwanida, N. (2015). A Review on the extraction methods use in medicinal plants, principle, strength and limitation. *Medicinal & Aromatic Plants, 04*(03). doi:10.4172/2167-0412.1000196

Backes, E., Pereira, C., Barros, L., Prieto, M. A., Genena, A. K., Barreiro, M. F., & Ferreira, I. C. F. R. (2018). Recovery of bioactive anthocyanin pigments from *Ficus carica* L. peel by heat, microwave, and ultrasound-based extraction techniques. *Food Research International, 113,* 197–209.

Charde, R. M., Dongade, H. J., Charde, M. S., & Kasture, A. V. (2010). Evaluation of antioxidant, wound healing and anti-inflammatory activity of ethanolic extract of leaves of *Ficus religiosa*. *International Journal of Pharmaceutical Sciences and Research, 1*(5), 73-85.

Coker, M. E., & Oaikhena, A. O. (2020). Antimicrobial Activity of the crude extracts and fractions of *Ficus thonningii* (Blume) on isolates from urinary tract infections. *Journal of Medicinally Active Plants, 9*(4), 310–322.

Eteraf-Oskouei, T., Allahyari, S., Akbarzadeh-Atashkhosrow, A., Delazar, A., Pashaii, M., Gan, S. H., & Najafi, M. (2015). Methanolic extract of *Ficus carica* Linn. leaves exert antiangiogenesis effects based on the rat air pouch model of inflammation. *Evidence-Based Complementary and Alternative Medicine,* 1-9.

Farnsworth, N. R. (1966). Biological and Phytochemical Screening of Plants. *Journal of Pharmaceutical Science, 55*(3), 225-276.

Fongang, Y. S. F., Bankeu, J. J. K., Ali, M. S., Awantu, A. F., Zeeshan, A., Assob, C. N., Mehreen, L., Lenta, B. N., Ngouela, S. A., & Tsamo, E. (2015). Flavonoids and other bioactive constituents from *Ficus thonningii* Blume (Moraceae). *Phytochemistry Letters,* 11, 139–145.

Ghimire, P., Pandey, S., Shrestha, A. C., & Chandra Shrestha, A. (2020). A comprehensive review on phytochemical, pharmacognostical properties and pharmacological activities of *Ficus lacor* L. (Moraceae). *International Journal of Herbal Medicine, 8*(5), 96–102.

Guaadaoui, A., Benaicha, S., Elmajdoub, N., Bellaoui, M., Hamal, A., (2014). What is a bioactive compound? A combined definition for a preliminary consensus. *International Journal of Food Science Nutrition, 3*, 174–179.

Hayouni, E.A., Abedrabba, M., Bouix, M. & Hamdi, M. (2007). The effects of solvents and extraction method on the phenolic contents and biological activities in vitro of Tunisian *Quercuscoccifera* L. and *Juniperus phoenica* L. fruit extracts. *Food Chemistry, 105*, 1126- 1134.

Huang, Y., Xiao, D., Burton-Freeman, B. M., & Edirisinghe, I. (2016). Chemical changes of bioactive phytochemicals during thermal processing. *Reference Module in Food Science.* https://doi.org/10.1016/b978-0-08-100596-5.03055-9

Jovanović, A. A., Đorđević, V. B., Zdunić, G. M., Pljevljakušić, D. S., Šavikin, K. P., Gođevac, D. M., & Bugarski, B. M. (2017). Optimization of the extraction process of polyphenols from *Thymus serpyllum* L. herb using maceration, heat- and ultrasound-assisted techniques. *Separation and Purification Technology, 179*, 369–380.

Khoo, L. W., Abas, F., Audrey Kow, S., Lee, M. T., Tan, C. P., Shaari, K., & Tham, C. L. (2018). A comprehensive review on phytochemistry and pharmacological activities of *Clinacanthus nutans* (Burm.F.) Lindau. *Evidence-Based Complementary and Alternative Medicine*, -39.

Lansky, E. P., Paavilainen, H. M., Pawlus, A. D., & Newman, R. A. (2008). *Ficus* spp. (fig): Ethnobotany and potential as anticancer and anti-inflammatory agents. *Journal of Ethnopharmacology, 119*(2), 195–213.

Li, H., Deng, Z., Wu, T., Liu, R., Loewen, S. & Tsao, R. (2012). Microwave-assisted extraction of phenolics with maximal antioxidant activities in tomatoes. *Food Chemistry, 130*, 928-936.

Li, Z., Yang, Y., Liu, M., Zhang, C., Shao, J., Hou, X., Tian, J., & Cui, Q. (2021). A comprehensive review on phytochemistry, bioactivities, toxicity studies, and clinical studies on *Ficus carica* Linn. leaves. *Biomedicine & Pharmacotherapy, 137*, 113397.

Mawa, S., Husain, K., & Jantan, I. (2013). *Ficus carica* L. (Moraceae): Phytochemistry, traditional uses and biological activities. *Evidence-based Complementary and Alternative Medicine, 2013*, 1-8.

Mohammed, A. H., & Al-Saddi, D. (2003). Factors Affecting the Extraction Process. *Iraqi Journal of Chemical and Petroleum Engineering, 4*(2), 5-13.

Mohd-Nasir, H., Asyikin, Z., Aziz, A., Peng, W. L., & Hamidah Mohd-Setapar, S. (2020). Comparative study on bioactive constituents and extraction procedure of Malaysian medicinal plants: A review. *Advances in Engineering Research, 200*, 17-29.

Mudi, S. Y., Muhammad, Musa, J., & Datti, Y. (2015). Phytochemical screening and antimicrobial activity of leaves and fruits extract of *Ficus sycomorus*. *Chemistry Search Journal, 6*(1), 62-67.

Oladosu, I. A., Zubair, M. F., Ali, M. S., & Olawore, N. O. (2009). Anticandidal activity of volatile compounds from the root bark of *Ficus exasperata* Vahl-Holl (Moraceae). *Journal of Essential Oil-Bearing Plants, 12*(5), 562–568.

Oluwasesan, B. M., Agbendeh, Z. M., & Jacob, A. G. (2013). Comparative studies of phytochemical screening of *Ficus sycomorus* linn stem bark extract and *Piliostigma thonningii* roots extract. *Pelagia Research Library Asian Journal of Plant Science and Research, 3*, 69–73.

Osorio-Tobón, J. F. (2020). Recent advances and comparisons of conventional and alternative extraction techniques of phenolic compounds. *Journal of Food Science and Technology, 57*(12), 4299-4315.

Panche, A. N., Diwan, A. D., & Chandra, S. R. (2016). Flavonoids: An overview. *Journal of Nutritional Science, 5*, 1-15.

Pereira, C. G., & Meireles, M. A. A. (2009). Supercritical fluid extraction of bioactive compounds: Fundamentals, applications and economic perspectives. *Food and Bioprocess Technology, 3*(3), 340–372.

Sapkale, G. N., Patil, S. M., Surwase, U. S., & Bhatbhage, P. K. (2010). Supercritical fluid extraction. *International Journal Chemistry Science, 8*(2), 729-743.

Sharma, M., & Sajgotra, M. (2017). Phytochemical screening and thin layer chromatography of *Ficus carica* leaves extract. *UK Journal of Pharmaceutical and Biosciences, 5*(1), 18-23.

Sharma, B. R., Kumar, V., Kumar, S., & Panesar, P. S. (2020). Microwave assisted extraction of phytochemicals from *Ficus racemosa. Current Research in Green and Sustainable Chemistry, 3*, 100020.

Sirisha, N., Sreenivasulu, M., Sangeeta, K., & Chetty, C. M. (2010). Antioxidant properties of *Ficus* species-A review. *International Journal of Pharmaceutical Technology Research, 2*(4), 2174-2182.

Supriya, J. G., Rohan, S. S., Neha, M. M., Minal, G. R., & Sanjay Smt Kashibai, S. D. (2012). Antimicrobial Activity Of *Ficus* glomerata Linn. Bark. *Interntional Research Journal of Pharmacy, 3*(5), 281-284.

Uku, U. P., Fokunang, T. E., Grace, M., Borgia, N. N., Mogue, I., Bathelemy, N., Nchafor, N. V., Fonmboh, J. D., Nyuki, A. B., Yves, T. O., Marcel, N. E., Tchaleu, N. B., & Ntungwen, F. C. (2020). Phytochemical screening and antiulcer activity, of *Ficus thonningii* (Moraceae) aqueous fruits extract in Wistar rats. *Asian Journal of Research in Medical and Pharmaceutical Sciences, 9*(1), 41–59.

Velavan, S. (2015). Phytochemical techniques-A review. *World Journal of Science and Research, 1*(2), 80-91.

Zhang, Q. W., Lin, L. G., & Ye, W. C. (2018). Techniques for extraction and isolation of natural products: A comprehensive review. *Chinese Medicine, 13*(20), 1-26.

Pharmacology of *Ficus*: A Review

Maizatul Akma Ibrahim[1,*] and Nor Hafizah Zakaria[1]
[1]Department of Plant Science, Kulliyyah of Science, International Islamic University Malaysia, Jalan Sultan Ahmad Shah, Bandar Indera Mahkota, 25200, Kuantan, Pahang.
Corresponding author: <u>maizatulakma@iium.edu.my</u>

ABSTRACT

Ficus is a medicinal plant that is commonly used in Malaysia, China, India, Thailand, South Africa, and other countries due to its pharmacological properties such as antimicrobial, anticancer, anti-inflammatory, antioxidant, analgesic, and antihypertensive activities. Based on the ethnological reports, this genus has been used to treat fever, skin diseases, hypertension, diabetes, cardiovascular diseases, and diarrhea. Women also have used the decoction of the whole plant as herbal drink to recover after childbirth. It also improves blood circulation and regains body strength, as well as treats disorders related to menstrual cycle. Recently, there were a lot of research about the isolation of the chemical constituents in the *Ficus* genus to understand the functions and the mechanisms of these constituents that attributed to the pharmacological properties of the plant. These chemical constituents, such as flavonoids, triterpenoids, steroids, and furocoumarins, have shown diverse applications in pharmacotherapeutic field. Therefore, this comprehensive review summarises the background and pharmacological properties of the *Ficus* genus.

Keywords: *Ficus*; pharmacology; phytochemicals; extract

INTRODUCTION

The genus *Ficus* of Moraceae family includes a large number of species, which is more than 800 species (Berg, 2003), mostly found in tropical areas of East Asia. The leaves are usually simple and waxy, and most exude white or yellow latex when broken. Many species have aerial roots, and a number are epiphytic. The unusual fruit structure, known as a syconium, is hollow, enclosing an inflorescence with tiny male and female flowers lining the inside. The growth forms of *Ficus* are varied, including shrubs, woody lianas, hemiepiphytes, epiphytes, and trees. Some indoor varieties of *Ficus,* such as fiddle-leaf fig and Audrey fig can be grown as outdoor ornamental plants as well (Al-Boudi and Afifi, 2011). *Ficus* species are rich in nutritional components and used as a source of food in Egypt, India, south China, Turkey, and Malaysia. Not all fruits from this genus can be eaten. For example, mistletoe fig or *F. deltoidea*, a tropical shrub that has rounded leaves and readily bears small fruits, unfortunately inedible. Only a handful with fruits considered edible such as *Ficus* carica L. or common fig, which is the only *Ficus* species that is cultivated for its fruit (Figure 1). This species is native to southwest Asia and the Mediterranean region. Other species like *F. erecta* often cultivated as pot plants in Taiwan, Indonesia and Japan (Aghel et al., 2009).

Fig. 1: Fruit of the common fig (*Ficus carica*) (Peter Firus, 2021)

ECOLOGY

Ficus trees are ecologically significant keystone species because they provide food and shelter to many seed-dispersing animals in tropical forest ecosystems. A large number of vertebrates feed on their fruits more than other plants. Shanahan et al. (2001) stated that figs are the most widely eaten fruit because approximately >10% of the world's birds and >6% of the world's mammals consume figs. The fruits are an important food resource to some frugivores, including fruit bats, capuchin monkeys, gibbons, and orangutans. Other than that, bird species like hornbill, Asian barbets, pigeons, and fig-parrot rely on fig fruits as their source of food. High levels of carbohydrate and calcium in figs make them attractive to frugivorous birds and mammals (Kinnaird and O'Brien, 2005). Fig wasp plays an important role in the pollination system in tropical forest ecology. The fruit ripens quickly after the pollen-bearing wasp leaves a *Ficus* plant, providing a rich feast that attracts a host of mammals and birds. If the plants were to be cut out of a forest or the fig wasps were removed, there would certainly be a dramatic change in the animal population, as reported by the reduction of population densities of fruit-eating mammals on small islands that lack *Ficus* species (Shanahan et al., 2001). The taxonomy of *Ficus* is described in Table 1.

Table 1: Taxonomy of *Ficus* (Gupta and Singh, 2012)

Domain	Eukaryota
Kingdom	Plantae
Subkingdom	Viridaeplantae
Phylum	Tracheophyta
Subphylum	Euphyllophytina
Division	Magnoliophyte
Class	Magnoliopsida
Subclass	Dilleniidae
Order	Urticales
Family	Moracaea
Tribe	Ficeae
Genus	Ficus

Phytochemical constituents

Ficus genus is reported to be responsible for the treatment of various disease conditions because it contains a number of biologically active compounds, which are also known as phytochemicals. These bioactive compounds are naturally expressed by the plant in response to biotic and abiotic stresses (Chawla et al., 2012). The major phytochemicals such as flavonoids and phenolic compounds are reported to be concentrated in the leaves, fruit pulp, roots, stem bark or wood, peel and seeds of different species of *Ficus* plant along with polyphenol, polysterols and triterpenoids (Mandal et al., 1999). Fig fruit, especially the skin or exocarp and seeds, consist of monosaccharide sugars and a mix of phytochemicals such as gallic acid, flavonoids, rutin, chlorogenic acid and epicatechin. Various pigments like polyphenols, flavonoids, and anthocyanins are responsible for the various colors shown in different species of *Ficus* (Ahmed and Urooj, 2010).

Table 2. List of *Ficus* species

Species name	Locality	Phytochemical compound	Part used	Pharmacological properties
F. abutilifolia	Nigeria	Tannins, saponins and flavanoids	Leaves	Antimicrobial activity (Taiwo et al., 2016)
F. afzelii	Egypt	Tannins, flavonoids and phenolics	Pulp and leaves	Antioxidant activity (Abdel-Hameed, 2009)
F. amplissima	India	Sterols, phenolic acids and triterpenoids	Bark and fruit	Antidiabetic and antioxidant activities (Arunachalam and Parimelazhagan, 2013)
F. arnottiana	India	Alkaloids, glycosides, saponins, flavonoids, phenolics and carbohydrates.	Leaves	Mucoprotective activity and gastric secretory (Babu et al., 2017)
F. asperifolia	Africa	Flavonoids, alkaloids, tannins and phenolic acids	Roots and stem	Antioxidant, antiulcer, anticancer activities (Okwu, 2004)
F. auriculata	China	Phenolic acids, flavonols and sterols	Fruit and leaves	Hepatoprotective and antidiabetic activities (El-Fishawy et al., 2011)
F. beecheyana	Taiwan	Sterols and gallic acids	Rhizome and roots	Antidiabetic, antioxidant and anticancer properties (Yen et al., 2017)
F. benghalensis	India, Pakistan and Nepal	Amino acids, pigments, furocoumarins, steroids and triterpenes	Bark, fruit, root and whole	Antioxidant, hypolipidemic, antibacterial, anti inflammatory, analgelsic and anticancer activities (Aswar et al., 2008)
F. benjamina	Hawaii and Australia	Steroids, flavonoids and phenolic acids	Leaves	Antioxidant, antimicrobial and antidiarrheal activities (Jain et al., 2013)
F. capensis	Tropical Africa, Cape Island	Polyphenols	Stem bark and leaves	Antibacterial, antiulcer, antidiarrheal, immune-boosting properties (Oyeleke et al., 2008)

F. caprefolia	South Africa	Phenolics	Leaves and latex	Antidiabetic activity (Olaokun et al., 2013)
F. carica	Asia, South America and Europe	Furocoumarins, flavonoids, phenolic acids and coumarins	Leaves, latex, fruit and roots	Hepatoprotective, laxative, antidysenteric activities (Jeong et al., 2009)
F. chlamydocarpa	Cameroon	Triterpenoids and flavonoids	Stem bark	Antimicrobial, hepatoprotective and antioxidant activities (Donfack et al., 2010)
F. cordata	Egypt, Saudi Arabia and Africa	Flavonoids, coumarins, phenolics and terpenoids	Leaves and stem bark	Antioxidant activity (Ahmed et al., 2017)
F. craterostoma	South Africa	Phenolics	Leaves	Antibacterial (Oyeleke et al., 2008)
F. crocata	Mexico	Triterpenoids and sterols	Leaves	Anticancer, antioxidant and analgelsic activities (Sánchez-Valdeolívar et al., 2020)
F. decora	Egypt	Phenolics, flavonoids and tannins	Leave	Antioxidant activity (Abdel-Hameed, 2009)
F. dekdekena	Africa and Senegal	Phenolics	Roots and leaves	Antioxidant activity (Olaokun et al., 2013)
F. deltoidea	Malaysia and Indonesia	Triterpenoids, terpenoids and flavonoids	Roots and leaves	Antidiabetic, anti inflammatory, anticancer, antibacterial (Zakaria et al., 2012)
F. drupacea	Vietnam	Triterpenoids and steroids	Stem bark, leaves	Antimalaria, anticancer, antimicrobial activities (Yessoufou et al., 2016)
F. elastica	India	Phenolic, flavonoids and tannins	Leaves	Antioxidant and antimicrobial activities (Preeti et al., 2015)
F. exasperata	Asia and South Africa	Phenolics and tannins	Leaves	Antidiabetic, anticonvulsant, antiinflammatory, antimicrobial, hypolipidemic, antioxidant and antiulcer activities (Ahmed et al., 2012)

F. foveolata	Thailand and Pakistan	Coumarins, flavonoids and triterpenoids	Stem bark	Antimicrobial activity (Meerungrueang et al., 2014)
F. glumosa	Ivory Coast and Central African	Flavanoid, alkaloids and tannins	Roots and stem bark	Hypolipidemic, anti inflammatory and hyperglicemic activities (Abu et al., 2020)
F. hirta	China	Furocoumarins, steroids and triterpenoids	Roots	Anticancer and antifungal properties (Wan et al., 2017)
F. hispida	India, Malaysia	Flavonoids, sterols and phenols	Leaves, stem bark and roots	Antiulcerogenic, cardioprotective and antidiabetic activties (Ali and Chaudhary, 2012)
F. ingens	Zimbabwe, Nigeria and Southen Arabia	Tannins and phenols	Stem bark and leaves	Analgelsic and antimicrobial activities (Olayinka et al., 2017)
F. insipida	Bolivia, Amazon	Triterpenoids	Leaves and fruits	Antianaemic and antipyretic activities (Gonzales et al., 2019)
F. lacor	India	Steroids and triterpenoids	Roots	Antiarthritic activity (Sindhu and Arora, 2013)
F. lutea	South Africa	Phenolics	Leaves	Antioxidant activity (Olaokun et al., 2013)
F. lyrata	Egypt	Flavonoids, tannins and phenolics	Leaves	Antibacterial and antioxidant activities (Abdel-Hameed, 2009)
F. mollis	India	Triterpenoids and flavonoids	Leaves and stem bark	Hypoglicemic and hypolipidemic activities (Munna and Saleem, 2013)
F. mysorensis	Egypt	Sterols and triterpenoids	Leaves	Antiinflammatory and anticancer activities (Abbass et al., 2015)
F. natalensis	Uganda and Nigeria	Flavonoids, alkaloids, saponins and steroids	Leaves	Antimicrobial activity (Sheyin et al., 2018)
F. nitida	Egypt	Flavonoids, tannins and phenolics	Leaves	Antioxidant activity (Abdel-Hameed, 2009)

F. nymphaefolia	Brazil	Isoflavones	Stem bark and latex	Wound healing (Darbour et al., 2007)
F. oligodon	China	Flavonoids and phenolics	Leaves	Antioxidant activity (Shi et al., 2011)
F. palmata	Mid-Himalayan region, Somalia and Sudan	Sterols, anthocyanins, flavonoids, furocoumarins and terpenoids	Stem bark, leave and roots	Antimicrobial, hepatoprotective and antiulcer activities (Joshi et al., 2014)
F. pandurata	Egypt	Steroids and triterpenes	Leaves	Analgelsic and antipyretic activities (Khedr et al., 2015)
F. platyphylla	Nigeria	Saponins and tannins	Stem bark	Antimicrobial and anticonvulsant properties (Kubmawara et al., 2009)
F. polita	Sourthen Africa	Phenolic acids and antocyanins	Leaves and roots	Antiviral, antimalaria and antimicrobial activities (Kuete et al., 2011)
F. pomifera	India	Phenols, tannins and triterpenoids	Leaves	Anticancer property (Wangkheirakpam et al., 2015)
F. pumila	South china and Malaysia	Steroids and flavonoids	Stem and leaves	Analgelsic and anti inflammatory activities (Liao et al., 2012)
F. racemosa	India, Pakistan, Sri Lanka	Steroids and triterpenoids	Whole plant	Hypoglicemic, anti cancer, hepatoprotective activities (Ahmed and Urooj, 2010)
F. religiosa	Nepal, Pakistan and India	Alkaloids, polyphenols, furocoumarin, steroids and triterpenoids	Leave, fruit and bark	Cardioprotective, antidiabetic, antitumor, antioxidant, antihelmintic, antimicrobial and antiparasitic activities (Damanpreet and Rajesh, 2009)
F. retusa	Peninsular Malaysia and India	Polyphenols	Leaves and stem barks	Hepatoprotective properties (Jaya Raju and Sreekanth, 2011)

Species	Origin	Phytochemicals	Part used	Activity (Reference)
F. semicordata	Iraq, India, Bangladesh and Myanmar	Tannins, alkaloids and steroids	Leaves, fruit and latex	Antidiarrheal and antioxidant activities (Gupta and Acharya, 2019)
F. septica	Papua new guinea	Alkaloids and tannins	Leaves	Antimicrobial activity (Damu et al., 2005)
F. sur	Somalia, Yemen and Nigeria	Saponins, tannins and phenols	Stem bark	Anticonvulsant activity (Ishola et al., 2013)
F. sycomorus	Israel and Zimbabwe	Flavonoids, phenolic acids and sterols	Whole plant	Antibacterial activity (Mohamed et al., 2010)
F. thonningii	Nigeria, Senegal, Angola and Mali	Alkaloids, terpenoids, flavonoids and tannins	Leaves, stem bark and roots	Antimicrobial and antidiarrheal activities (Dangarembizi et al., 2012)
F. tikoua	China	Phenolics and isoflavonoids	Stem bark	Anti-diarrheal, antioxidant and antifungal activities (Jiang et al., 2013)
F. tinctoria	India	Flavonoids, alkaloids, glycosides, tannins and saponins	Whole plant	Antioxidant, antiulcer and antidiabetic properties (Gini *et al.,* 2017)
F. tsiela	India	Alkalaoids flavanoids coumarins saponins and terpenoids	Leaves	Antipneumonia and antimicrobial activities (Vaya and Mahmood, 2006).
F. ulmifolia	Philippine	Steroids, terpenoids and sterols	Leaves	Antioxidant activity (Ragasa et al., 2009)
F. umbelatte	Africa	Coumarins	Stem bark	Menapousal problem (Zingue et al., 2016)
F. vallis-choudae	Ivory Coast and Cameroon	Triterpenoids and sterols	Leaves, stem	Anticonvulsant, antifungal and anti inflammatory properties (Bnakeu et al., 2017)
F. virens	Pakistan and Egypt	Phenolics and flavonoids	Bark, latex and leaves	Antioxidant, anti diabetic properties (Orabi and Orabi, 2016)

Pharmacological properties

Due to their great pharmacological benefits, *Ficus* species have been described by many authors in both traditional and modern medicines.

Antioxidant activity

The presence of flavonoids in *F. carica* leaves could be responsible for the antioxidant activity, as reported by Ali et al. (2012). In another report by Konyalıoğlu et al. (2015), antioxidant capacity based on TAC method revealed tocopherol equivalents/g dry mass ranged from 14.04±1.42 to 23.51±1.15 mM in *F. carica* leaves extract of methyl alcohol, ethyl alcohol, *n*-hexane, ethyl acetate, and water. The highest antioxidant activity was evident from the water extract of *F. carica* leaves. Using well-established modified DPPH (2,2-diphenyl-1-picrylhydrazyl) method, the highest polarity of crude extract of *F. carica* and *F. sycomorous* detected a high content of flavonoids and phenol compounds which contributed to the antioxidant activity in *Ficus* (Al-Matani et al., 2015; Weli et al., 2015). In the research conducted by Kaur et al. (2010), the DPPH test revealed antioxidant activity ranging from 6.34 to 13.35% for *F. religiosa* in different concentrations of ethanolic extract of the leaf (200-1000g/mL). In a comparison of methanolic stem bark and root extracts of *F. glomerata,* the investigation revealed high antioxidant activity in the stem bark extract (Channabasavaraj et al., 2008).

Another research done by Jahan et al. (2009), antioxidant efficacy of *F. racemosa* fruits extract was proven using in vitro assay of DPPH free radical scavenging capacities. Five ethanolic extracts of *Ficus* species, namely *F. auriculata, F. virens, F. callosa* and *F. vasculosa* were tested for antioxidant efficacy via *in vitro* assay. Both *F. auriculata* and *F. virens* have shown IC_{50} values of 0.3 mg/mL, which were higher than other *Ficus* extracts, making them the most promising species for antioxidant activity (Shi et al., 2011). Meanwhile, Omar et al. (2011) investigated the flavonoids and phenolic compounds extracted from the leaf of *F. deltoidea*. The chromatographic method on a reversed-phase C_{12} column revealed a high total of flavonoid content. The result of the HPLC method exhibited that 85% of total antioxidant activity was produced by the leaf extract.

Antimicrobial activity

F. religiosa has shown antimicrobial activity against a various type of pathogenic microorganisms (Salehi et al., 2020). In other research, a minimum inhibitory concentration (MIC) was tested in several bacteria namely, *Shigella flexneri, Enterococcus faecalis, Shigella dysenteriae, Proteus vulgaris, Shigella sonnie* and *Staphylococcus saprophyticus* using ethanolic extract of *F. religiosa*. The result showed that the extract can inhibit these bacterial with the MIC value ranged from 250 to 500μg/mL (Rahman et al., 2014). Antibacterial activity of the ethanolic fruit extract from *F. religiosa* also has been done in India. The researchers reported that this extract had shown an inhibition against *S. aureus* and *S. epidermidis* with MIC of 15 mg/mL. This extract also showed inhibition against *K. pneumoniae* and *P. vulgaris* with an effective MIC of 30 mg/mL (Kumar Goyal et al., 2014).

The methanolic extract and a compound, namely ficusoside isolated from the root of *F. elastica*. Ficusoside showed a MIC of 4.9 μg/mL, which lower than gentamicin and fluconazole (25 μg/mL) against *S. aureus, E. coli, P. vulgaris* and *C. albicans*. In comparison, the methanolic extract showed a MIC of 39.1 μg/mL (Mbosso et al., 2017). Acetone leaf and stem bark extract of *F. sycomorous* showed a good antibacterial activity by inhibiting resistant *A. baumannii.* The leaf extract recorded a MIC of 2.5 mg/mL, whereas stem bark showed 4.9 mg/mL of MIC (Saleh et al., 2015). Manimozhi et al. (2012) reported the antibacterial activity of the flavonoids extracted from the bark of *F. bengalensis.* The evaluation showed that the flavonoids are excellent in inhibiting the growth of *P. vulgaris, S. aureus* and *P. aeruginosa* with MIC ranging from 25 to 100 mg/mL. Evaluation of antibacterial activities of methanol bark extract of *F. microcarpa* showed that the extract was effective against the tested gram-negative and gram-positive bacteria. The result of inhibition zones against *B. brevis, B. cereus, B. subtilis, E. coli* and *A. polymorph* were 18.0, 15.5, 16.5, 16.0 and 8.0 mm, respectively (Ao et al., 2008).

Wound healing activity

The healing property of the root extract of *F. racemosa* in albino rats was evaluated. In aqueous and ethanolic root extract of treated groups, the application of extract increased the breaking strength of the incision wounds was which may be due to the presence of an individual or combined action of phytochemicals such as saponins, flavonoids, alkaloids and tannins (Murti and Kumar, 2012). In other studies, 5% and 10% of hydroalcoholic leaf extract of *F. religiosa* has exhibited a healing process in the incision wound, excision wound and burn wound induced to the rats (Naira et al., 2009). Aqueous extract of *F. racemosa* was found to possess a wound healing property. When 10% dose of the extract was applied to the wound, the healing process was accelerated (Mehta et al., 2012).

The wound-healing efficacy of ethanolic and aqueous extracts of *F. benghalensis* was evaluated in albino rats (Garg and Paliwal, 2011). It was evident that the 200 mg/kg b.w. dose of the extract showed a healing activity by the improvement in the wound region. In the wound healing experiment, the aqueous extract of *F. deltoidea* was used in induced wounds of rats. The evaluation has shown that the wounds applied with 5% and 10% of *F. deltoidea* extract significantly promoted the rate of wound healing compared to wounds treated with sterile deionized water or dressed with blank placebo (Abdulla et al., 2010). Murti et al. (2011) also investigated the healing activity of *F. bengalensis* in albino rats. The healing process in the excision, incision and dead space wound was accelerated with the treatment of the aqueous and ethanolic root extract. The extracts showed that the period of epithelialization reduced, the breaking strength increased, the rate of wound contraction increased and the hydroxyproline content elevated.

Hypoglycemic activity
β-Sitosterol-D-glycoside was isolated from the root bark of *F. religiosa* and *F. glomerata*. This compound could be responsible for hypoglycemic activity in *Ficus*. The concentration of 25, 50 and 100 mg/kg of dose were administered orally in induced diabetic rats and normal rats. A significant decrease in blood glucose was observed using the dose of 50 and 10mg/kg compared to 25mg/kg (Chandrasekar et al., 2010). In combination with *F. racemosa* extract and hypoglycemic drug, hypoglycemic activity was studied in diabetic patients. The extract and the drug were taken orally for 15 days. The result showed that the blood glucose level was significantly reduced. The renal and liver functions were also tested to rule out the herb toxicity, which was observed to be in the normal range (Gul-e-Rana et al., 2012).

Isolated flavonoids from the stem bark of *F. racemosa* were evaluated for antidiabetic activity. At 100mg/kg dose, the flavonoids were administered to streptozotocin (STZ) rats. The level of blood glucose was measured on different days (1st, 3rd, 5th and 7th days). The finding showed that the flavonoids able to reduce the blood glucose level, which could be useful as a supplementary drug for future diabetic therapy (Keshari et al., 2016). The methanolic extract of the bark of *F. amplissima* was tested for hypoglycemic activity in streptozotocin-induced diabetic rats. At 50mg/kg and 100mg/kg of dose, the extract has exhibited a reduction in total cholesterol, blood glucose level and serum triglyceride and (Arunachalam et al., 2013). In other research conducted by Gayathri and Kannabiran, (2008), an aqueous extract from the bark of *F. benghalensis* was evaluated in diabetic rats. The result showed that the extract significantly reduces glucose levels. The level of hepatic cytochrome P450 dependent enzyme, blood electrolytes and systems glycolytic enzymes were also restored.

Hyperlipidemic activity
The hyperlipidemic activity of ethanolic extract of *F. racemosa* bark in alloxan-induced diabetic rats was investigated. The finding showed that 300 mg/kg dose able to restore the level of lipids and lipoproteins to near normal range (Sophia and Manoharan, (2007). This showed that this extract could be a potent supplementary drug in combination with the standard reference drug, glibenclamide. In the research of dexamethasone-induced hyperlipidemia in rats, ethyl acetate leaf extract of *F. mollis* was tested using two different doses, 200 and 400 mg/kg. The extract reverted the hyperlipidemia caused by dexamethasone in a dose-dependent manner. The extract showed a similar effect with the reference standard, glibenclamide (Munna and Saleem, 2013). In the experiment of Triton WR 1339-induced hyperlipidemia in swiss albino mice, leaf and twig extract of *F. carica* were evaluated. After administration of doses 150 and 300 mg/kg, the hyperlipidemic effect was examined by the various parameter of lipid profile. The result showed that the extracts cause a reduction in the level of serum total cholesterol, triglycerides and low-density lipoprotein (LDL) (Boukhalfa et al., 2018). In other research, isolated α-amyrin acetate from aerial roots of

F. bengalensis was administered orally to db/db mice for ten days. At 50 mg/kg dose, the extract shows a reduction in triglycerides, cholesterol and LDL-C by 21.5%, 24.1% and 21.2%, respectively (Singh et al., 2009).

Anti-inflammatory effect

Ethanolic extract of *F. religiosa* leaf showed a significant anti-inflammatory property in the study conducted by Charde et al. (2010). The edema was induced in Wistar rat using carrageenan, and the extract was administered topically. At a concentration of 300 μg/mL, the edema decreased as the extract could be inhibiting the release of serotonin, histamine, kinins and prostaglandins. The effect was similar with the application of ibuprofen as a control. An aqueous extract from the leaves of *F. racemosa* was tested for the anti-inflammatory property on serotonin, histamine, carrageenan and dextran-induced hind paw edema model in Wistar rats. The extract (400mg/kg) showed maximum inhibition of 32%, 34%, 30%, and 31% in serotonin, histamine, carrageenan, and dextran-induced rats, respectively. A similar effect was also observed with the standard drug, phenylbutazone (Mandal et al., 2000).

In other studies, the anti-inflammatory effect of chloroform, petroleum ether and ethanol extracts from *F. carica* the leaves was studied on a carrageenan-induced rat paw edema model. The ethanol extract (600mg/kg) showed a maximum anti-inflammatory effect (76%) in acute inflammation and granuloma weight has exhibited a decrease of 72% (Patil and Patil, 2011). The anti-inflammatory effect of ethanolic bark extract of *F. bengalensis* was better than petroleum ether. Carrageenan was used to induce hind paw edema in rats and the oral administration was done with the dose of 300 and 600 mg/kg. Both ethanolic extracts for 300 and 600 mg/kg have shown an inhibitory effect against carrageenan-induced edema, which indicated that these extracts possess an anti-inflammatory effect for acute inflammation (Patil et al., 2009).

Hepatoprotective activity

The hepatoprotective activity of methanolic and petroleum ether of *F. racemosa* stem bark was tested in CCl4-induced hepatic damage in rats. The administration of CCl4 reduced the level of albumin, serum total protein and urea and increased the total bilirubin associated with an increment in alanine aminotransferase (ALT), aspartate aminotransferase (AST) and alkaline phosphatase (ALP) activities as compared to control rats. Both methanol and petroleum ether extracts restored albumin and total protein to near normal levels. Both extracts also reduced the level of ALT, AST and ALP. Total bilirubin content also reduced from 2.1 mg/dL to 0.3 mg/dL. These results showed that *F. racemosa* possesses potent hepatoprotective effects against CCl4-induced hepatic damage in rats.

The hepatoprotective effect of *F. religiosa* latex on cisplatin-induced liver injury in Wistar rats was investigated. The increase of serum ALP, ALT, AST and hepatocytes cells degeneration inflammatory infiltrate and necrosis were due to the cisplatin-induced liver impairment. The methanolic extract of *F. religiosa* latex restored serum ALP, ALT, AST, lipid peroxidation, SOD and GSH of the liver to near normal levels. Generally, *F. religiosa* latex is effective in protection and improvement against cisplatin-induced liver injury (Yadav, 2015). Leaf ethanolic extract of *F. carica* was tested in CCl4-induced hepatic damage in albino rats. Liver markers and histopathological changes were examined. The result showed that the dose of the extract (200 mg/kg) had improved the hepatic damaged by CCl4 (Aghel et al., 2011). Joshi et al. (2014) also reported the hepatoprotective effect of total extract of *F. palmata* in animals with hepatic toxicity. The elevated AST, ALT, GGT, ALP and bilirubin have exhibited a significant reduction at a dose of 400 mg/kg body weight.

Antiulcer activity

The objectives of the antiulcer treatment are to accelerate the healing process, improve the symptoms, prevent ulcer recurrence, and eradicate the presence of *H. pylori*. In the study of antiulcer activity, stomach lesion was induced in the rats with the treatment of ethanol. The total extract of *F. palmata* was used at 200 and 400 mg/kg against 80% of ethanol-induced stomach lesion. The result showed that the best protection against ulcer was achieved by the highest dose (400 mg/kg), where the ulcer index was 2.00 ± 0.57 (Alqasoumi et al., 2014). In other works, hydroalcoholic and ethanolic of *F. religiosa* of leaves and stem bark extract were tested for antiulcer activity in Sprague–Dawley rats with gastric ulcer induced by aspirin, ethanol, and ligated pylorus (Saha and Goswami, 2010). A significant reduction in ulcer index can be seen when applying 250 and 500 mg/kg body weight of extract at the ulcer area. This effect was equivalent to the standard drug reference, antacid.

Antiulcer activity of *F. religiosa* ethanolic leaf extract was evaluated in albino mice. Treatment with the extract has shown a significantly reduced gastric lesion formation and submucosal edema similar to the ranitidine-treated mice. The result also showed that there was no sign of toxicity and mortality at the high dose of 2 000 mg/kg, which indicated that the extract was safe and non-toxic even at high concentrations (Gregory et al., 2013). According to Sivaraman and Muralidhara, (2010), the application of *F. hispida* extract has reduced gastric ulceration in rats. An investigation of antiulcer efficacy has been done on methanolic extract of *F. hispida* with doses of 200 and 400 mg/kg. These doses were found to be effective in healing the ulcer by 64% (200mg/kg) and 69% (400mg/kg) and reducing free and total acidity as well.

Anticancer activity
Some species of the *Ficus* genus show the anticancer property in cell lines of several cancer types. Hexane, dichloromethane and acetone extract of *F. crocata* have been evaluated for antiproliferative activity in cell line of breast cancer. Dichloromethane extract showed the strongest effect in reducing the proliferation of MDA-MB-231 cells (Sánchez-Valdeolívar et al., 2020). *F. carica* also showed anticancer activity when the methanolic extract of leaf and fruit were evaluated against Huh7it liver cancer cells using MTT assay (Purnamasari et al., 2019). The result showed that the extracts had IC_{50} values >653 µg/mL for the leaf extract and >2000 µg/mL for the fruit extract. A higher percentage of Huh7it apoptosis and necrosis in the leaf compared with fruit extracts was also observed.

In another work by Bunawan et al. (2014), the anticancer activity of ethanolic and aqueous extracts of *F. deltoidea* was tested against human ovarian carcinoma cell line A2780 using MTT assay. The ethanolic and aqueous extracts gave IC_{50} values of 143.03±20.21 µg/mL and 224.39+6.24 µg/mL, respectively. Both extracts also showed an apoptosis at 1000 µg/mL. It can be found that the ethanolic extract reduced cell proliferation, while aqueous extract induced cell detachment. Various phytochemical contents in the extracts could be attributed to this finding.

Antidiarrheal activity
Mandal and Kumar (2002) investigated the antidiarrheal activity in leaves extract of *F. hispida*. Diarrhea in rats was induced with castor oil and PEG_2. The result showed that the methanol leaves extract could be used as an antidiarrheal agent as it inhibited the activities of diarrhea and enteropooling in rats. It was assumed that tannins might be responsible for the antidiarrheal activity as it denatures the protein and forms protein tannate, which minimizes the intestinal mucosa permeability. Meanwhile, the leaf extract of *F. microcarpa* has been used in investigating the effect of antidiarrheal activity in rats. Castor oil was used to induce diarrhea in rats. The oral administration of the extract at doses of 300 and 600 mg/kg produced a significant antidiarrheal effect in rats. At 300 mg/kg, the percentage of inhibition based on the number and weight of feces was 79% and 66%. While, at 600 mg/kg, the values for both number and weight were 32%. Based on the volume and weight of intestinal content, there was also a reduction in anti-enteropooling activity (Bairagi et al., 2014).

Cardioprotective effect
Leaf extract of *F. hispida* was prepared to investigate the cardioprotective effect on cyclophosphamide mediated myocardial injury due to oxidative stress in rat heart (Shanmugarajan et al., 2008). This finding revealed that lipid peroxidation was inhibited significantly. There was also an increased concentration of glutathione reductase, superoxide dismutase, glutathione peroxidase, catalase, and glutathione-S-

transferase. Glutathione activity in heart tissue also decreased caused by cyclophosphamide. Its cardioprotective activity could be due to antioxidant constituents, which could be responsible for this finding such as hispidin, β-sitosterol, β-amyrin, and bergaptin.

The extract of *F. religiosa* showed an improvement in oxidative stress, diabetic markers, and inflammatory and cardiac markers in streptozotocin-induced diabetic cardiomyopathy rats. The control of cytokine, diabetes and modulation of oxidative marker could be attributed to the cardioprotective role of *F. religiosa* (Singh et al., 2011). The extract of *F. thonningii* was prepared to investigate the myocardial contractile performance on rat isolated atrial muscle strips. The result showed that there were negative inotropic and chronotropic effects on both spontaneously beating and electronically driven atrial muscle strips (Musabayane et al., 2007). According to Baur and Sinclair, (2006), the resveratrol in *F. thonningii* could be attributed to the cardioprotective effects of *F. thonningii*. Other than cardiovascular diseases, resveratrol is also responsible for delaying the aging process and prevent the progression of various diseases, which include obesity, cancers and neurodegenerative disorders (Ramawat et al., 2009).

CONCLUSION

This review covers various pharmacological properties of *Ficus* spp., in vitro and in vivo trials including antiproliferative, antioxidant, antimicrobial and anti-inflammatory activities. The review also exposed the discovery and isolation of the plant metabolites such as sterols, flavonoids, terpenoids, saponins, coumarins and alkaloids, which could be contributed to the therapeutic potential of the plant. This plant can be considered safe and non-toxic as no severe side effects were reported.

REFERENCES

Abbass, H. S., Ragab, E. A., Mohammed, A. E. I. & El-Hela, A. A. (2015). Investigation of *Ficus mysorensis* cultivated in Egypt. *Journal of Pharmaceutical, Chemical and Biological, 3*(3), 396-407.

Abdel-Hameed, E. S. S. (2009). Total phenolic contents and free radical scavenging activity of certain Egyptian Ficus species leaf samples. *Food Chemistry, 114*(4), 1271-1277.

Abdulla, M. A., Ahmed, K. A., Abu-Luhoom, F. M. & Muhanid, M. (**2010**). Role of *Ficus deltoidea* extract in the enhancement of wound healing in experimental rats. *Biomedical Research, 21*(3), 241-245.

Abu, M. S., Yakubu, O. E., Boyi, R. N., Mayel, M. H. & Haibu, C.O. (2020). Phytochemical constituents, acute toxicity and free radical scavenging activity of methanol extract of *Ficus glumosa* leaves. *Anchor University Journal of Science and Technology, 1*(1), 16-22.

Aghel, N., Kalantari, H. & Rezazadeh, S. (2011). Hepatoprotective Effect of *Ficus carica* leaf extract on mice intoxicated with carbon tetrachloride. *Iranian Journal of Pharmaceutical Research, 10* (1), 63-68.

Ahmed, F. A., Mohamed, M. A., Abdel-Aziem, A. & El-Azab, M. M. (2016). Phytochemical composition of *Ficus cordata* Thunb. subsp. salicifolia (Vahl) and its antioxidant activity with lead induced testicular toxicity in rats. *International Journal of Innovative Science, Engineering & Technology, 4*(10), 64-82.

Ahmed, F., Mueen-Ahmed, K. K., Abedin, M. Z. & Karim, A. A. (2012). Traditional uses and pharmacological potential of *Ficus exasperata* Vahl. *Systematic Reviews in Pharmacy, 3*(1), 15-23.

Ahmed, F. & Urooj, A. (2010). Traditional uses, medicinal properties, and phytopharmacology of *Ficus racemosa*: A review. *Pharmaceutical Biology, 48*(6), 672-681.

Al-Aboudi, A. & Afifi, F. U. (2011). Plants used for the treatment of diabetes in Jordan: A review of scientific evidence. *Pharmaceutical Biology, 49*(3), 221-239.

Al-Matani, S. K., Al-Wahaibi, R. N. S. & Hossain, M. A. (2015). In vitro evaluation of the total phenolic and flavonoid contents and the antimicrobial and cytotoxicity activities of crude fruit extract with different polarities from *Ficus sycomorus*. *Pacific Science Review A: Natural Science and Engineering, 17*(3), 103–108.

Ali, M. & Chaudhary, N. (2011). *Ficus hispida* Linn.: A review of its pharmacognostic and ethnomedicinal properties. *Pharmacognosy reviews, 5*(9), 96–102.

Alqasoumi, S. I., Basudan, O. A., Al-Rehaily, A. J. & Abdel-Kader, M. S. (2014). Phytochemical and pharmacological study of *Ficus palmata* growing in Saudi Arabia. *Saudi Pharmaceutical Journal, 22*(5), 460–471.

Arunachalam, K. & Parimelazhagan, T. (2013). Antidiabetic activity of *Ficus amplissima* Smith. bark extract in streptozotocin induced diabetic rats. *Journal of Ethnopharmacology, 147*(2), 302-310.

Aswar, M., *Aswar, U., Watkar, B., Vyas, M., Wagh, A. & Gujar, K. N.* (2008). Anthelmintic activity of *Ficus benghalensis. International Journal of Green Pharmacy, 2*(3), 170-172.

Babu, A., Anand, D. & Saravanan, P. (2017). Phytochemical analysis of *Ficus arnottiana* (Miq.) Miq. leaf extract using GC-MS analysis. *International Journal of Pharmacognosy and Phytochemical Research, 9*(6), 775-779.

Bairagi, S. M., Aher, A. A, Nema, N. & Pathan I. B. (2014). Evaluation of anti-diarrhoeal activity of the leaves extract of *Ficus microcarpa* L. (Moraceae). *Marmara Pharmaceutical Journal, 18*, 135-138.

Bankeu, J. J. K., Dawé, A., Mbiantcha, M., Feuya, G. R. T., *et al.,* (2017). Characterization of bioactive compounds from *Ficus vallis-choudae* Delile (Moraceae). *Trends in Phytochemical Research, 1*(4), 235-242.

Baur, J. A. & Sinclair, D. A. (2006). Therapeutic potential of resveratrol: The *in vivo* evidence. *Nature Reviews Drug Discoveries, 5*, 493-506.

Berg, C. C. (2003). Flora Malesiana precursor for the treatment of Moraceae 1: The Main subdivision of Ficus: The subgenera. *Blumea-Biodiversity, Evolution and Biogeography of Plants, 48*(1), 166–177.

Boukhalfa, F., Kadri, N., Bouchemel, S., Ait Cheikh, S., Chebout, I., Madani, K. & Chibane, M. (2018). Antioxidant activity and hypolipidemic effect of *Ficus carica* leaf and twig extracts in Triton WR-1339-induced hyperlipidemic mice. *Mediterranean Journal of Nutrition and Metabolism, 11*(1), 37-50.

Bunawan, H., Mat Amin, N., Bunawan, S. N., Baharum, S. N. & Mohd Noor, N. (2014). *Ficus deltoidea* Jack: A review on its phytochemical and pharmacological importance. *Evidence-Based Complementary and Alternative Medicine.*

Cagno, V., Civra, A., Kumar, R., Pradhan, S., Donalisio, M., Sinha, B. N., Ghosh, M. & Lembo, D. (2015). *Ficus religiosa* L. bark extracts inhibit human rhinovirus and respiratory syncytial virus infection in vitro. *Journal of Ethnopharmacology, 24*(176), 252-257.

Chandrasekar, S. B., Bhanumathy, M., Pawar, A. T. & Somasundaram, T. (2010). Phytopharmacology of *Ficus religiosa. Pharmacognosy reviews, 4*(8), 195–199.

Changwei, A. O., Li, A., Elzaawely, A. A., Xuan, T. D. & Tawata, S. (2008) Evaluation of antioxidant and antibacterial activities of *Ficus microcarpa* L. fil. Extract. *Food Control, 19*(10), 940-948.

Channabasavaraj, K. P., Badami, S. & Bhojraj, S. (2008). Hepatoprotective and antioxidant activity of methanol extract of *Ficus glomerata. Journal of Natural Medicines, 62*(3), 379–383.

Chawla, A., Kaur, R. & Sharma, A.K. (2012). *Ficus carica* Linn, a review on its pharamacognostic, phytochemical and pharmacologica aspects. *International Journal of Pharmaceutical and Phytopharmacology Research, 1*, 215–32.

Dangarembizi, R., Erlwanger, K. H., Moyo, D. & Chivandi, E. (2012). Phytochemistry, pharmacology and ethnomedicinal uses of *Ficus thonningii* (Blume Moraceae): a review. *African Journal of Traditional and Complementary Alternative Medicine, 10*(2), 203-212.

Damanpreet, S. & Rajesh, K. G. (2009). Anti-convulsant effect of *Ficus religiosa*: role of serotonergic pathways. *Journal of Ethnopharmacology, 123*(2), 330-334.

Damu, A. G., Kuo, P. C., Shi, L. S., Li, C. Y., Kuoh, C. S. Wu, P. L. & Wu, T. S. (2005). Phenanthroindolizidine alkaloids from the stems of *Ficus septica*. Natural Product, 68(7), 1071–1075.

Darbour, N., Bayet, C., Rodin-Bercion, S., Elkhomsi, Z., Felix Lurel, F., Chaboud, A. & David Guilet, D. (2007). Isoflavones from *Ficus nymphaefolia. Natural Product Research, 21*(5), 461-464.

Donfack, I H., Simo, C. C. F., Ngameni, B., Tchana, A. N., *et al.* (2010). Antihepatotoxic and antioxidant activities of methanol extract and isolated compounds from *Ficus chlamydocarpa. Natural Product Communications, 5*(10), 1607-1612.

El-Fishawy, A., Zayed, R. & Afifi, S. (2011). Phytochemical and pharmacological studies of *Ficus auriculata* Lour. *Journal of Natural Products. 4*,184-195.

Garg, V. K., & Paliwal, S. K. (2011). Wound-healing activity of ethanolic and aqueous extracts of *Ficus benghalensis. Journal of Advanced Pharmaceutical and Technology Research, 2*(2), 110-114.

Gini, E. J., Sivakkumar, T. & Kuppuswami, S. (2017). Determination of antioxidant activity of different extracts of *Ficus gibbosa* Blume, isolation and characterization of flavonoid from ethanol extract by column chromatography. *Research Journal of Pharmaceutical, Biological and Chemical Sciences, 8*(4), 888-898.

Gonzales, A. P. P. F., Santos, G. G. & Tavares-Dias, M. (2019). Anthelminthic potential of the *Ficus insipida* latex on monogeneans of *Colossoma macropomum* (Serrasalmidae), a medicinal plant from the Amazon. *Acta Parasitology, 64*, 927–931.

Gregory, M., Divya, B., Mary, R. A., Viji, M. M. H., Kalaichelvan, V. K. & Palanivel, V. (2013). Anti-ulcer activity of *Ficus religiosa* leaf ethanolic extract. *Asian Pacific Journal of Tropical* Biomedicine, 3(7), 554–556.

Gul-e-Rana, Karim, S., Khurhsid, R., Saeed-ul-Hassan, S., Tariq, I., Sultana, M., Rashid, A. J., Shah, S. H. & Murtaza, G. (2013). Hypoglycemic activity of *Ficus racemosa* bark in combination with oral hypoglycemic drug in diabetic human. *Acta Poloniae Pharmaceutica-Drug Research, 70*(6), 1045-1049.

Gupta, S. & Acharya, R. (2019). Antioxidant and nutritional evaluation of *Bhu Udumbara* (*Ficus semicordata* Buch.-Ham. ex Sm.) leaves and fruits: An extra pharmacopoeial drug of Ayurveda. *Ayu, 40*(2), 120–126.

Gupta, C. & Singh, S. (2012). Taxonomy, phytochemical composition and pharmacological prospectus of *Ficus religiosa* linn. (Moraceae)- A review. *Journal of Phytopharmacology, 1*(1), 57-70.

Ishola, I. O., Olayemi, S. O., Yemitan, O. K. & Ekpemandudiri, N. K. (2013). Mechanisms of anticonvulsant and sedative actions of the ethanolic stem-bark extract of *Ficus sur* Forssk (Moraceae) in rodents. *Pakistan Journal of Biological Science, 16*(21), 1287-94.

Jahan, I. A., Nahar, N., Mosihuzzaman, M., Rokeya, B., Ali, L., Azad Khan, A. K. & Iqbal Choudhary, M. (2009). Hypoglycaemic and antioxidant activities of *Ficus racemosa* Linn. Fruits. *Natural Product Research, 23*(4), 399–408.

Jain, A., Ojha, V., Kumar, G., Karthik, L., Venkata, K. & Rao, B. (2013). Phytochemical composition and antioxidant activity of methanolic extract of *Ficus benjamina* (Moraceae) Leaves. *Research Journal of Pharmaceutical and Technology* 6(11), 1184-1189.

Jaya Raju, N. & Sreekanth, J. (2011). Investigation of hepatoprotective activity of *Ficus retusa* (moraceae). *International Journal of Research in Ayurveda and Pharmacy, 2* (1), 166-169.

Jeong, M. R., Kim, H. Y. & Cha, J. D. (2009). Antimicrobial activity of methanol extract from *Ficus carica* leaves against oral bacteria. *Journal of Bacteriology and Virology. 39*(2), 97-102.

Jiang, Z. Y., Li, S. Y., Li, W. J., Guo, J. M., Tian, K., Hu, Q. F. & Huang, X .Z. (2013). Phenolic glycosides from *Ficus tikoua* and their cytotoxic activities. *Carbohydrate Research, 15*(382), 19-24.

Kaur, H., Singh, D., Singh, B. & Goel, R. K. (2010). Anti-amnesic effect of *Ficus religiosa* in scopolamine-induced anterograde and retrograde amnesia. *Pharmaceutical Biology, 48*(2), 234–240.

Keshari, A. K., Kumar, G., Kushwaha, P. S., Bhardwaj, M., Kumar, P., Rawat, A., Kumar, D., Prakash, A., Ghosh, B. & Saha, S. (2016). Isolated flavonoids from *Ficus racemosa* stem bark possess antidiabetic, hypolipidemic and protective effects in albino Wistar rats. *Journal of Ethnopharmacology. 181*, 252-62.

Khedr, A. I. M., Allam, A. E., Nafady, A. M., Ahmad, A. S. & Ramadan, M. A. (2015). Phytochemical and biological screening of the leaves of *Ficus pandurata* Hance. cultivated in Egypt. *Journal of Pharmacognosy and Phytochemistry, 3*(6), 50-54.

Kinnaird, M. F. & O'brien, T.G. (2005). Fast Foods of the Forest: The Influence of Figs on Primates and Hornbills Across Wallace's Line. In: Dew J. L., Boubli J. P. (eds) Tropical Fruits and Frugivores. Springer, Dordrecht. 155-184.

Konyalıoğlu, S., Sağlam, H. and Kıvçak, B. (2005). α-Tocopherol, Flavonoid, and phenol contents and antioxidant activity of *Ficus carica*. Leaves, Pharmaceutical Biology. 43:8: 683-686.

Kubmarawa, D., Khan, M. E., Punah, A. M. & Hassan, M. (2009). Phytochemical and antimicrobial screening of *Ficus platyphylla* against human/animal pathogens. *The Pacific Journal of Science and Technology, 10*(1), 382-386.

Kuete, V., Kamga, J., Sandjo, L. P. *et al.* (2011). Antimicrobial activities of the methanol extract, fractions and compounds from *Ficus polita* Vahl. (Moraceae). *BMC Complementary Alternative Medicine, 11*(6), 1-6.

Kumar Goyal, A., Sharma, R., Kaur, R. & Kaushik, D. (2014). In vitro studies on antibiotic activity of *Ficus religiosa* fruits extract against human pathogenic Bacteria. *Journal of Chemical and Pharmaceutical Research, 6*(11), 80–84.

Liao, C., Kao, C., Peng, W., Chang, Y., Lai, S. & Ho, Y. (2012). Analgesic and anti-inflammatory activities of methanol extract of *Ficus pumila* L. in Mice. *Evidence-Based Complementary and Alternative Medicine.*

Mandal, S. C., Maity, T. K., Das, J., Saba, B. P. & Pal, M. (2000). Anti- inflammatory evaluation of *Ficus racemosa* Linn. Leaf extract. *Journal of Ethnopharmacology, 72*(1–2), 87–92.

Mandal, S. C. & Kumar, C. K. (2002). Studies on antidiarrhoeal activity of *Ficus hispida* leaf axtract in rats. *Fitoterapia. 73*, 663–667.

Manimozhi, D. M., Sankaranarayanan, S. & Sampathkumar, G. (2012). Evaluating the antibacterial activity of flavonoids extracted from *Ficus benghalensis*. *International Journal of Pharmaceutical and Biological Research, 3*(1), 7–18.

Mbosso, T. J. E. Noundou, X. S., Fannang, S., Meyer, F., Vardamides, J. C., Mpondo, E., Mpondo, R. W. M., Krause, A. G. B. & Azebaze, J. C. A. (2017). In vitro antimicrobial activity of the methanol extract and compounds from the wood of *Ficus elastica* Roxb. ex Hornem. aerial roots. *South African Journal of Botany, 111*, 302-306.

Meerungrueang, W. & Panichayupakaranant, P. (2014). Antimicrobial activities of some Thai traditional medical longevity formulations from plants and antibacterial compounds from *Ficus foveolata*. *Pharmaceutical Biology, 52*(9),1104-1109.

Mehta, D. S., Kataria, B. C. & Chhaiya, S. B. (2012). Wound healing and anti-inflammatory activity of extract of *Ficus racemosa* linn. bark in albino rats. *International Journal of Basic & Clinical Pharmacology, 1*(2), 111-115.

Mohamed, A. E. H. H., El-Sayed, M. A., Hegazy, M. E., Helaly, S. E., Esmail, A. M. & Mohamed, N. S. (2010). Chemical constituents and biological activities of Artemisia herba-alba. *Records of Natural Products. 4*(1), 1-25.

Munna, S. & Saleem, M. T. (2013). Hypoglycemic and hypolipidemic activity of *Ficus mollis* leaves. *Brazillan Journal of Pharmacognosy, 23*, 687-691.

Murti, K. & Kumar, U. (2012). Enhancement of wound healing with roots of *Ficus racemosa* L. in albino rats. *Asian Pacific Journal of Tropical Biomedicine, 2*(4), 276–280.

Musabayane, C. T., Gondwe, M., Kadyamaapa, D. R., Chuturgoon, A. A. & Ojewole, J. A. O. (2007). Effects of *F. thonningii* (Blume) Moraceae stem bark ethanolic extract on blood glucose, cardiovascular and kidney cell lines of the proximal (LLC-PK1) and distal tubules (MDBK). *Renal Failure, 29*, 389-397.

Naira, N., Rohini, R. M., Syed, M. B. & Amit, K. D. (2009). Wound healing activity of the hydro alcoholic extract of *Ficus religiosa* leaves in rats. *International Journal of Alternative Medicine, 6*, 2-7.

Okwu, D. E. (2004). Phytochemicals and vitamin content of indigenous species of South Eastern Nigeria. *Journal of Sustainable Agriculture and Environment, 6*, 30–34.

Olaokun, O. O. McGaw, L. J., Eloff, J. N. & Naidoo, V. (2013). Evaluation of the inhibition of carbohydrate hydrolysing enzymes, antioxidant activity and polyphenolic content of extracts of ten African Ficus species (Moraceae) used traditionally to treat diabetes. *BMC Complementary and Alternative Medicine, 13*(1), 94.

Olayinka, B. U., Abdulkareem, A. K., Adeyemi, S. B., Anwo, I. O., Lawal, A. R., Akinwunmi, M. A. & Etejere, E. O. (2017). *Ficus ingens* (Miq.) Miq. (Moraceae): phytochemical and proximate composition. *Annals of West University of Timişoara, ser. Biology. 20* (2), 153-158.

Omar, M. H., Mullen, W. & Crozier, A. (2011). Identification of pro- anthocyanidin dimers and trimers, flavone C-glycosides, and antioxidants in *Ficus deltoidea*, a Malaysian herbal tea. *Journal of Agricultural and Food Chemistry, 59*(4), 1363–1369.

Orabi, M. A. A. & Orabi, E. A. (2016). Antiviral and antioxidant activities of flavonoids of *Ficus virens*: Experimental and theoretical investigations. *Journal of Pharmacognosy and Phytochemistry. 5*(3), 120-128.

Oyeleke, S., Dauda, B., & Boye, O. (2008). Antibacterial activity of *Ficus capensis*. *African Journal of Biotechnology. 7*(10), 1414-1417.

Patil, V. V., Pimprikar, R. B. & Patil, V. R. (2009). Pharmacognostical studies and evaluation of anti-inflammatory activity of *Ficus bengalensis* Linn. *Journal of Young Pharmacists, 1*:49-53.

Peter, F. (2019). Fig. https://www.britannica.com/plant/fig (Accessed on 9 March 2021).

Purnamasari, R., Winarni, D., Permanasari, A. A., Agustina, E., Hayaza, S., & Darmanto, W. (2019). Anticancer activity of methanol extract of *Ficus carica* leaves and fruits against proliferation, apoptosis, and necrosis in Huh7it cells. *Cancer Informatics, 18*, 1-7.

Rahman, M., Khatun, A., Khan, S., Hossain, F. & Khan, A. (2014). Phytochemical, cytotoxic and antibacterial activity of two medicinal plants of Bangladesh. *Pharmacology, 1*(5), 3–10.

Ramawat, K. G., Doss, S. & Mathura, M. (2009). The chemical diversity of bioactive molecules and Therapeutic potential of medicinal plants. In: Ramawat, K.G., (eds.), Herbal Drugs Ethnomedicine to Modern Medicine, Springer-Verlag Berlin Heidelberg, pp. 7-31.

Saha, S. & Goswami, G. (2010). Study of antiulcer activity of *Ficus religiosa* L. on experimentally induced gastric ulcers in rats. *Asian Pacific Journal of Tropical Medicine, 3*(10), 791–793.

Sánchez-Valdeolívar, C. A., Alvarez-Fitz, P., Zacapala-Gómez, A. E. *et al.* (2020). Phytochemical profile and antiproliferative effect of *Ficus crocata* extracts on triple-negative breast cancer cells. *BMC Complementary Medicine and Therapy, 20*(1), 191.

Shanahan, M., So, S., Compton, S. G. & Corlett, R. T. (2001). Fig-eating by verterbrate frugivores: A global review. *Biological Reviews, 76*(4), 529-537.

Sheyin, F. T, Ndukwe, G. I, Iyun, O. R. A, Anyam, J. V. & Habila J. D. (2018). Phytochemical and antimicrobial screening of crude extracts of natal fig (*Ficus Natalensis* Kraus). *Journal of Applied Science and Environmental Management, 22*(9), 1457 –1460.

Shi, Y. X., Xu, Y. K., Hu, H. B., Na, Z. & Wang, W. H. (2011). Preliminary assessment of antioxidant activity of young edible leaves of seven Ficus species in the ethnic diet in Xishuangbanna, Southwest China. *Food Chemistry, 128*(4), 889-894.

Sindhu, R. K. & Arora, S. (2013). Therapeutic effect of *Ficus lacor* aerial roots of various fractions on adjuvant-induced arthritic rats. *International Scholarly Research Notices*.

Singh, D., Singh, B. & Goel, R. K. (2011). Traditional uses, phytochemistry and pharmacology of *Ficus religiosa*: a review. *Journal of Ethnopharmacology, 134*(3), 565-583.

Sivaraman, D. & Muralidharan, P. (2011). Anti-ulcerogenic evaluation of root extract of *Ficus hispida* linn: In aspirin ulcerated rats. *African Journal of Pharmaceutical and Pharmacology, 4*, 72–82.

Sophia, D. & Manoharan, S. (2007). Hypolipidemic activities of *Ficus racemosa* Linn. bark in alloxan induced diabetic rats. *African Journal of Traditional, Complementary, and Alternative Medicines, 4*(3), 279–288.

Taiwo, F. O., Fidelis, A. A. & Oyedeji, O. (2016). Antibacterial activity and phytochemical profile of leaf extracts of *Ficus abutilifolia*. *British Journal of Pharmaceutical Research. 11*(6), 1-10.

Vaya, J. & Mahmood, S. (2006). Flavanoid content in leaf extracts of the fig (*Ficus carica* L.) carob (*Ceratonia siliqua* L.) and Pistachio (*Pistacia lentiscus* L.), *Biofactors, 28*, 169– 75.

Wangkheirakpam, S. D., Wadawal, A., Leishangthem, S. S., Gurumayum, J. S. & Laitonjam, W. S. (2015). Cytotoxic triterpenoids from *Ficus pomifera* Wall. *Indian Journal of Chemistry. 54*, 676-681.

Wan, C., Chen, C., Li, M., Yang, Y., Chen, M. & Chen, J. (2017). Chemical constituents and antifungal activity of *Ficus hirta* Vahl. fruits. *Plants. 6*(4), 44.

Weli, A. M., Al-Blushi, A. A. M. & Hossain, M. A. (2015). Evaluation of antioxidant and antimicrobial potential of different leaves crude extracts of Omani *Ficus carica* against food borne pathogenic bacteria. *Asian Pacific Journal of Tropical Disease, 5*(1), 13–16.

Yadav, Y. C. (2015). Hepatoprotective effect of *Ficus religiosa* latex on cisplatin induced liver injury in Wistar rats. *Brazilian Journal of Pharmacognosy, 25*(3), 278-283.

Yen, G. C., Chen, C. S., Chang, W. T., Wu, M. F., Cheng, F. T., Shiau, D. K. & Hsu, C.L. (2017). Antioxidant activity and anticancer effect of ethanolic and aqueous extracts of the roots of *Ficus beecheyana* and their phenolic components. *Journal of Food and Drug Analysis*, 1-11.

Yessoufou, K., Elansary, H. O., Mahmoud, E. A. E. & Skalicka-Woźniak, K. (2016). Antifungal, antibacterial and anticancer activities of *Ficus drupacea* L. stem bark extract and biologically active isolated compounds. *Industrial Crops and Products, 74*, 752–758.

Zakaria, Z. A., Hussain, M. K., Mohamad, A. S., Abdullah, F. C. & Sulaiman, M. R. (2012). Anti-inflammatory activity of the aqueous extract of *Ficus deltoidea. Biological Research for Nursing, 14*(1), 90-7.

Zingue, S., Michel, T., Tchatchou, J., Chantal Beatrice Magne, N., Winter, E., Monchot, A., Awounfack, C. F., Djiogue, S., Clyne, C., Fernandez, X., Creczynski-Pasa, T. B. & Njamen, D. (2016). Estrogenic effects of *Ficus umbellata* Vahl. (Moraceae) extracts and their ability to alleviate some menopausal symptoms induced by ovariectomy in Wistar rats. *Journal of Ethnopharmacology, 17*(179), 332-44.

Printed by Books on Demand GmbH, Norderstedt / Germany